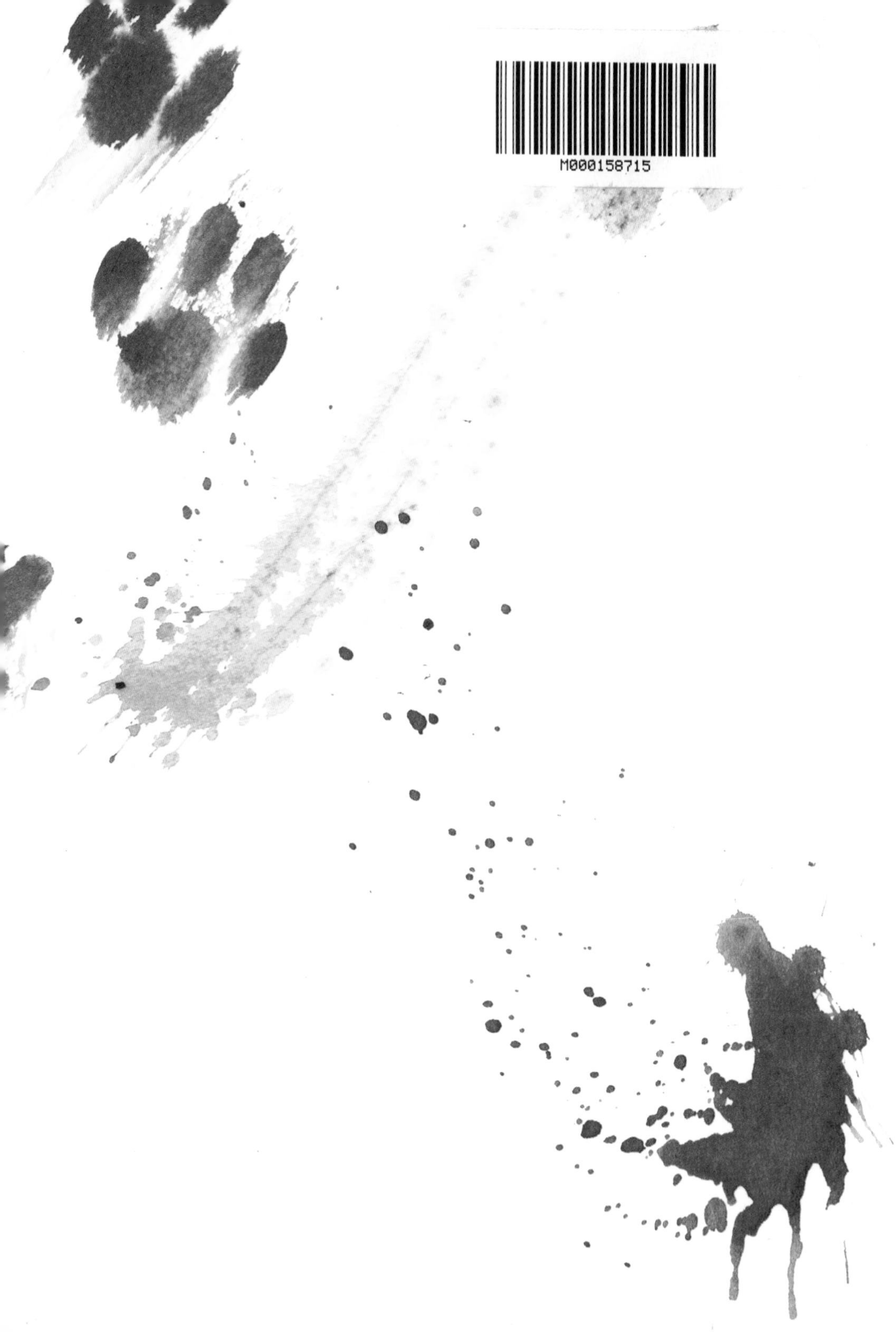

Keira & Me

Also by Noel Fitzpatrick

Becoming the Supervet: Listening to the Animals

Being the Supervet: How Animals Saved My Life

Beyond Supervet: How Animals
Make Us The Best We Can Be

For Children

Vetman and his Bionic Animal Clan

The Superpets and Me

NOEL FITZPATRICK

ILLUSTRATED BY LAURA McKENDRY

Keira & Me

SEVEN DIALS

First published in Great Britain in 2023 by Seven Dials
an imprint of The Orion Publishing Group Ltd
Carmelite House, 50 Victoria Embankment
London EC4Y 0DZ

An Hachette UK Company

5 7 9 10 8 6

A CIP catalogue record for this book is
available from the British Library.

ISBN (Hardback) 978 1 3996 1030 8
ISBN (eBook) 978 1 3996 1031 5

Designed by Hannah Beatrice Owens
Printed in Italy by L.E.G.O.

MIX
Paper | Supporting responsible forestry
FSC® C104740
FSC www.fsc.org

www.orionbooks.co.uk

To all who seek the warmth
of unconditional love,
May Keira's joy for truly living
illuminate your path,
as she did mine

Foreword

When I was a child, I wished on the brightest star in heaven that I could be stronger, braver and cleverer, so that I could save all the animals after I lost my first patient – a lamb in a moon-lit frosty field in Ireland.

That star was Sirius, which is the Dog Star and part of the constellation of *Canis Major*. Years later, while studying for further examinations, trying to become a specialist veterinary surgeon in neuro-orthopaedics, the light of my best friend, a Dog Star called Keira, sat at my feet and kept me going.

In being by my side through everything as I built my veterinary practice, Fitzpatrick Referrals, she lit the path for tens of thousands of animals we have treated through the years, and many thousands more who have been treated by the colleagues we have trained and inspired along the way. Keira in no small way was integral to all of this and all of them. Keira made me, and represents so much of what I made. She looked out for me and looked right through me, too. She saved me from myself and, in so doing, inspired the salvation of countless lives, including my own.

Keira wasn't always good; in fact, she was often naughty, in myriad funny ways. The half-chewed slipper, the bin liner torn apart, the tap turned on and the room flooded . . . But with her bristly eyes looking up at me, all was forgiven in an instant.

She taught me about the forgiveness of others, and of myself, too. She was truth and reconciliation all wrapped up in one boisterous bundle of bristles. I often felt I was not good enough, but she was always 'enough', even when she wasn't good.

Love,
Noel x

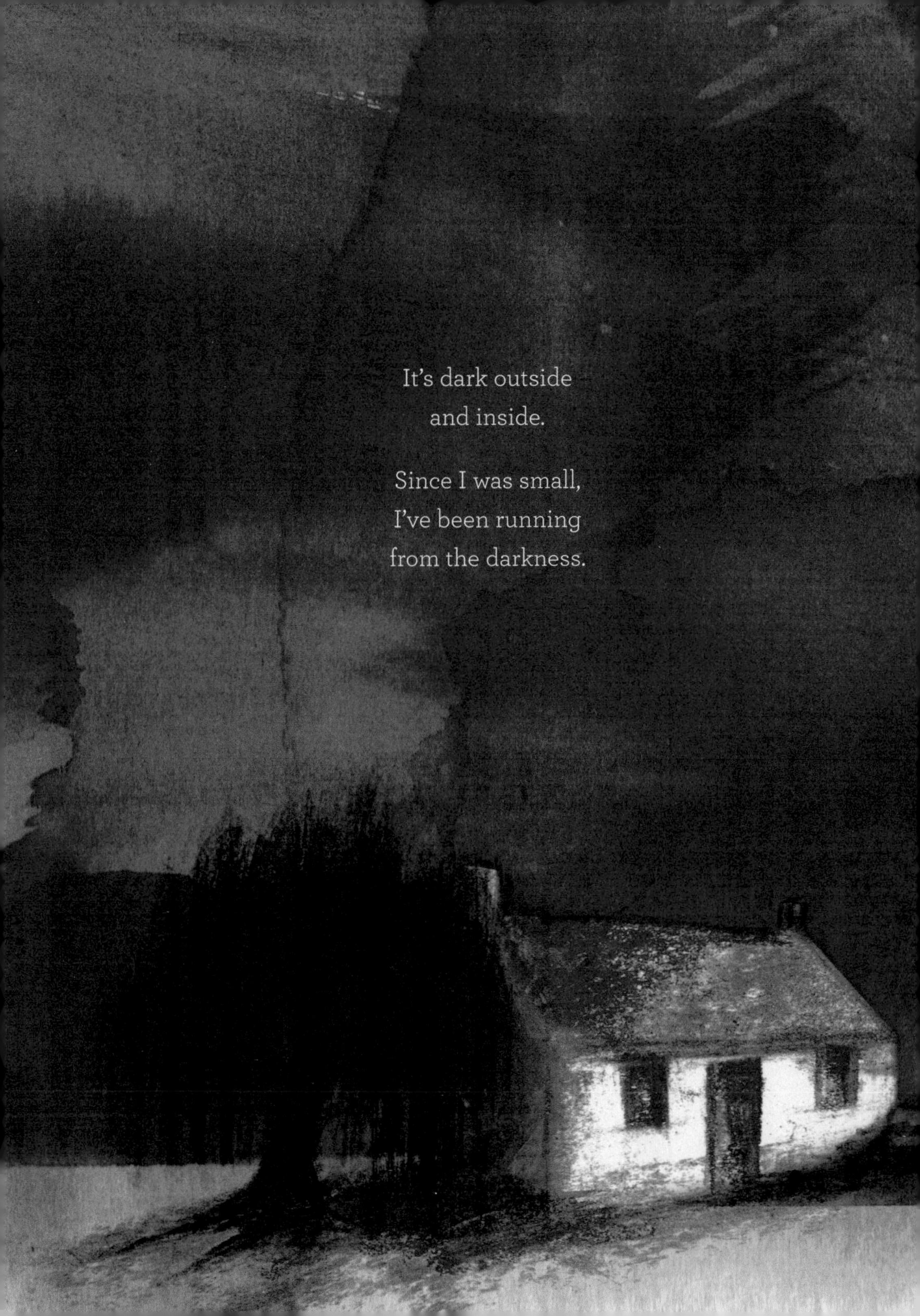

It's dark outside
and inside.

Since I was small,
I've been running
from the darkness.

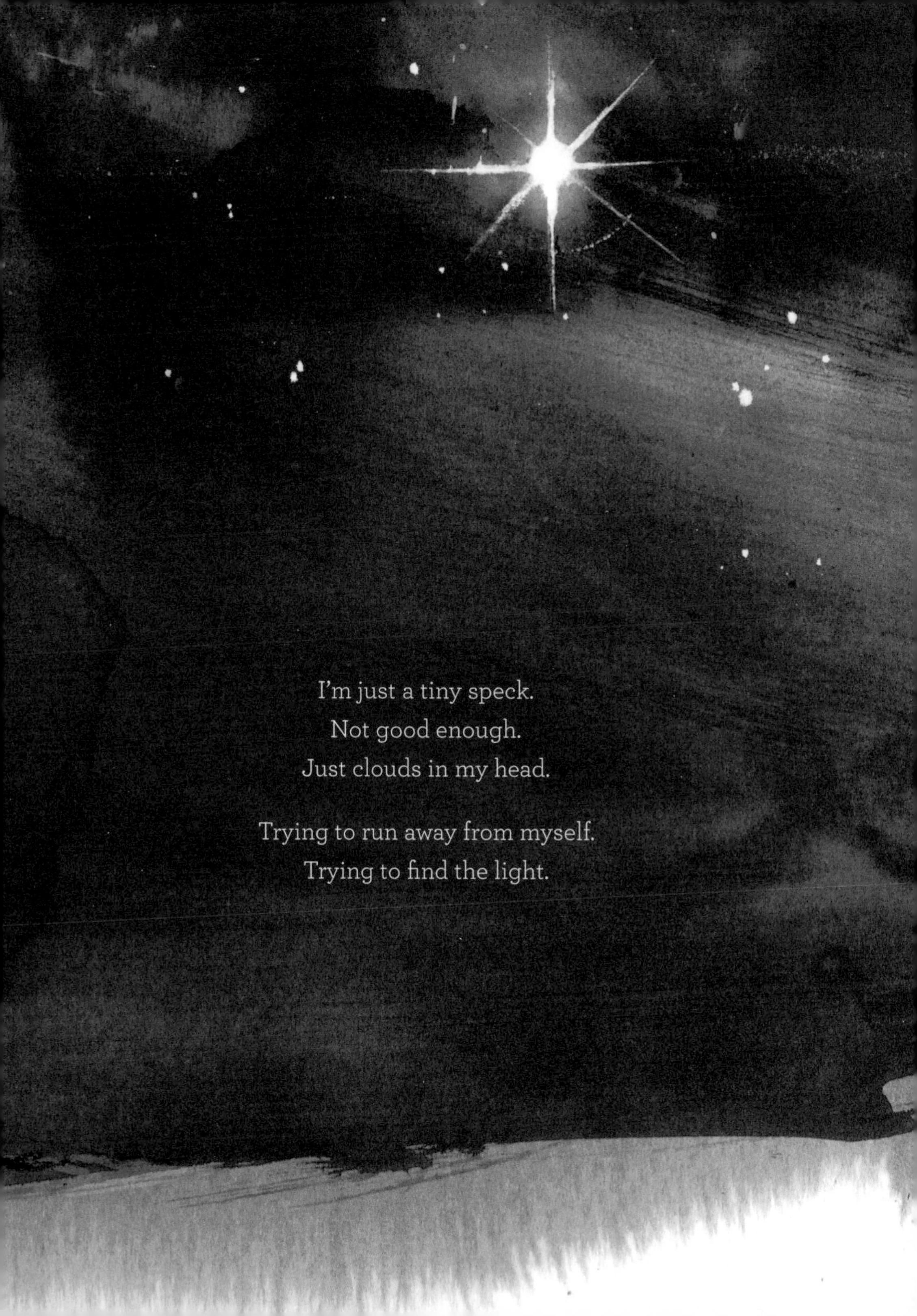

I'm just a tiny speck.
Not good enough.
Just clouds in my head.

Trying to run away from myself.
Trying to find the light.

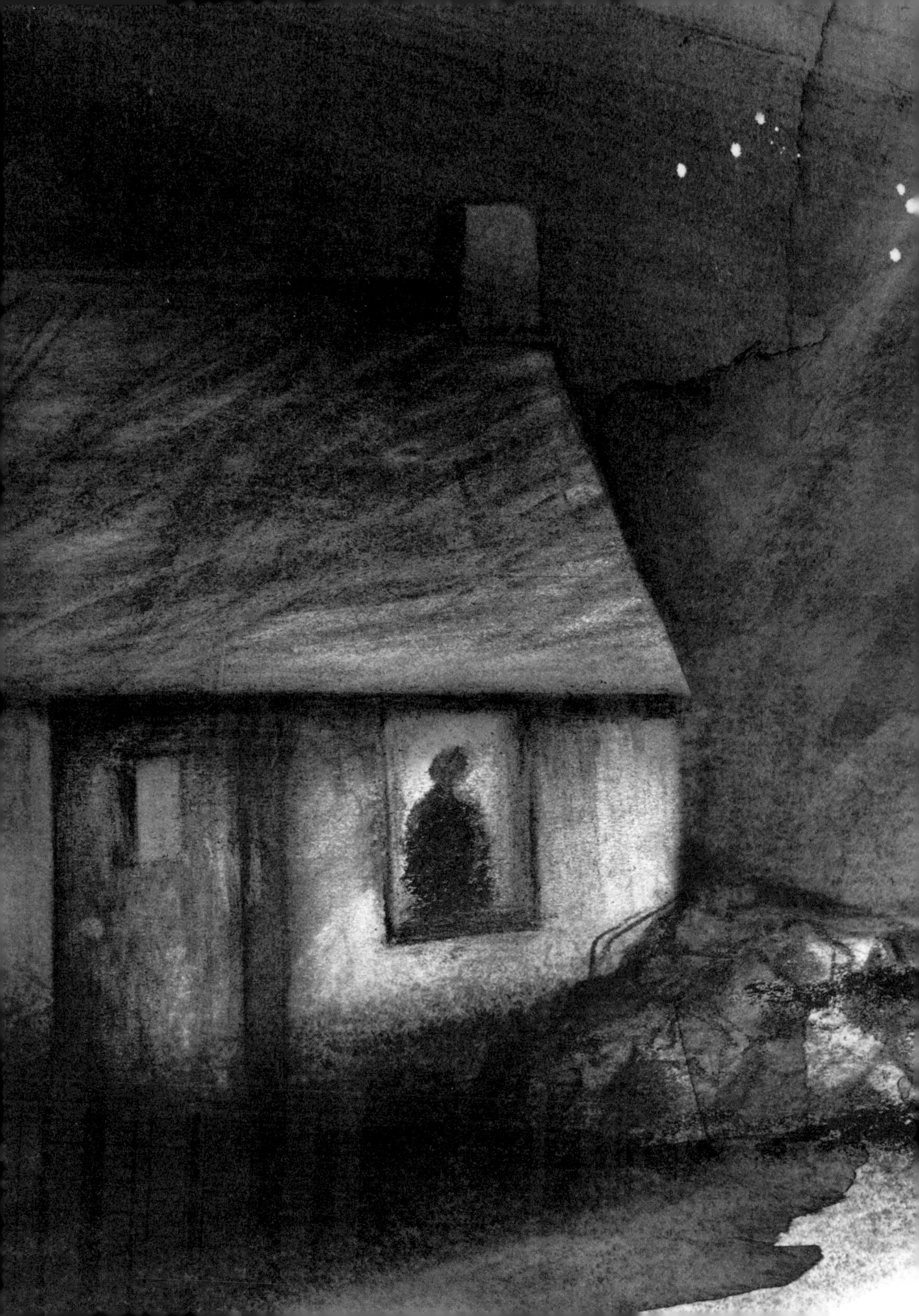

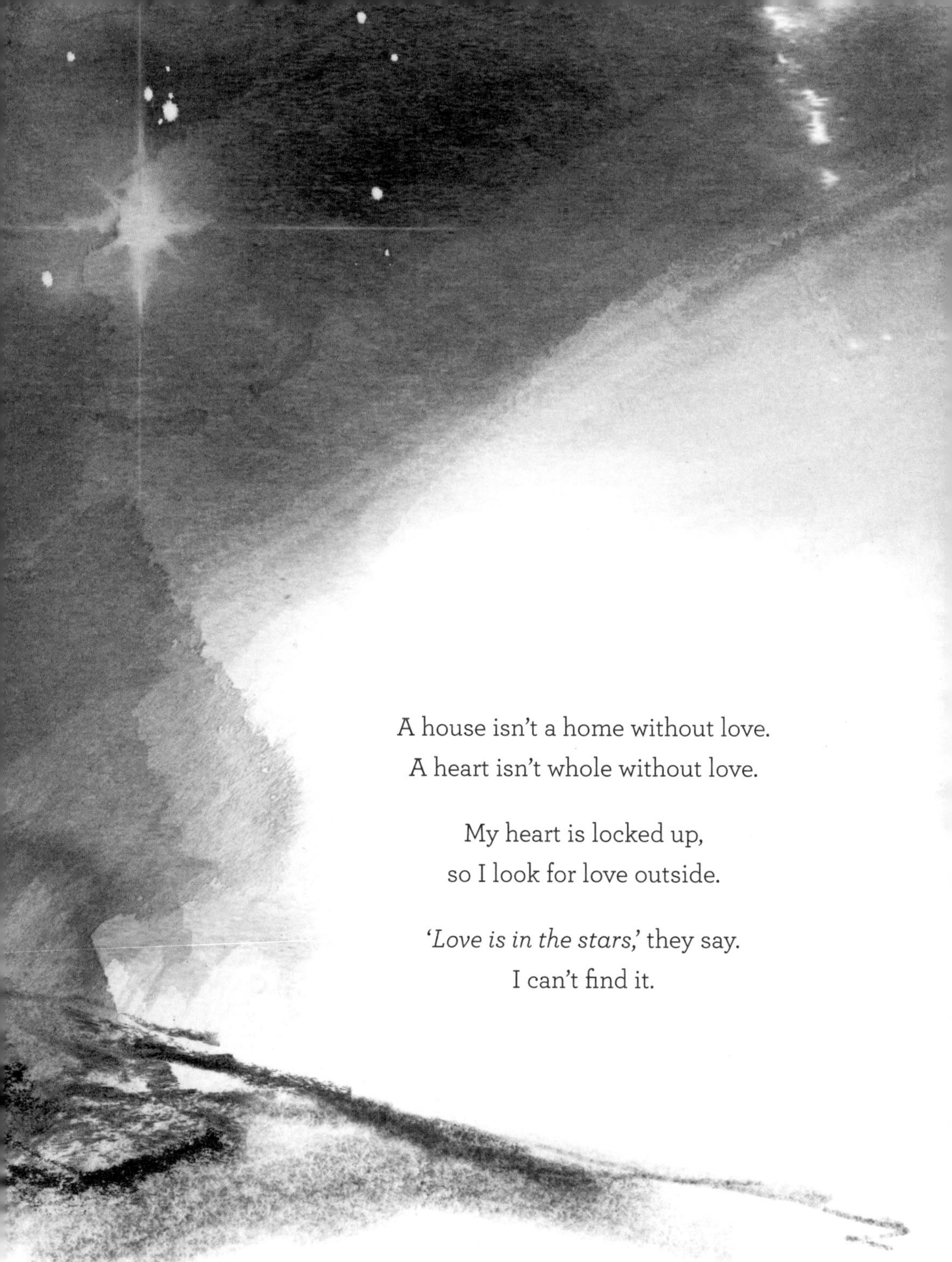

A house isn't a home without love.
A heart isn't whole without love.

My heart is locked up,
so I look for love outside.

'Love is in the stars,' they say.
I can't find it.

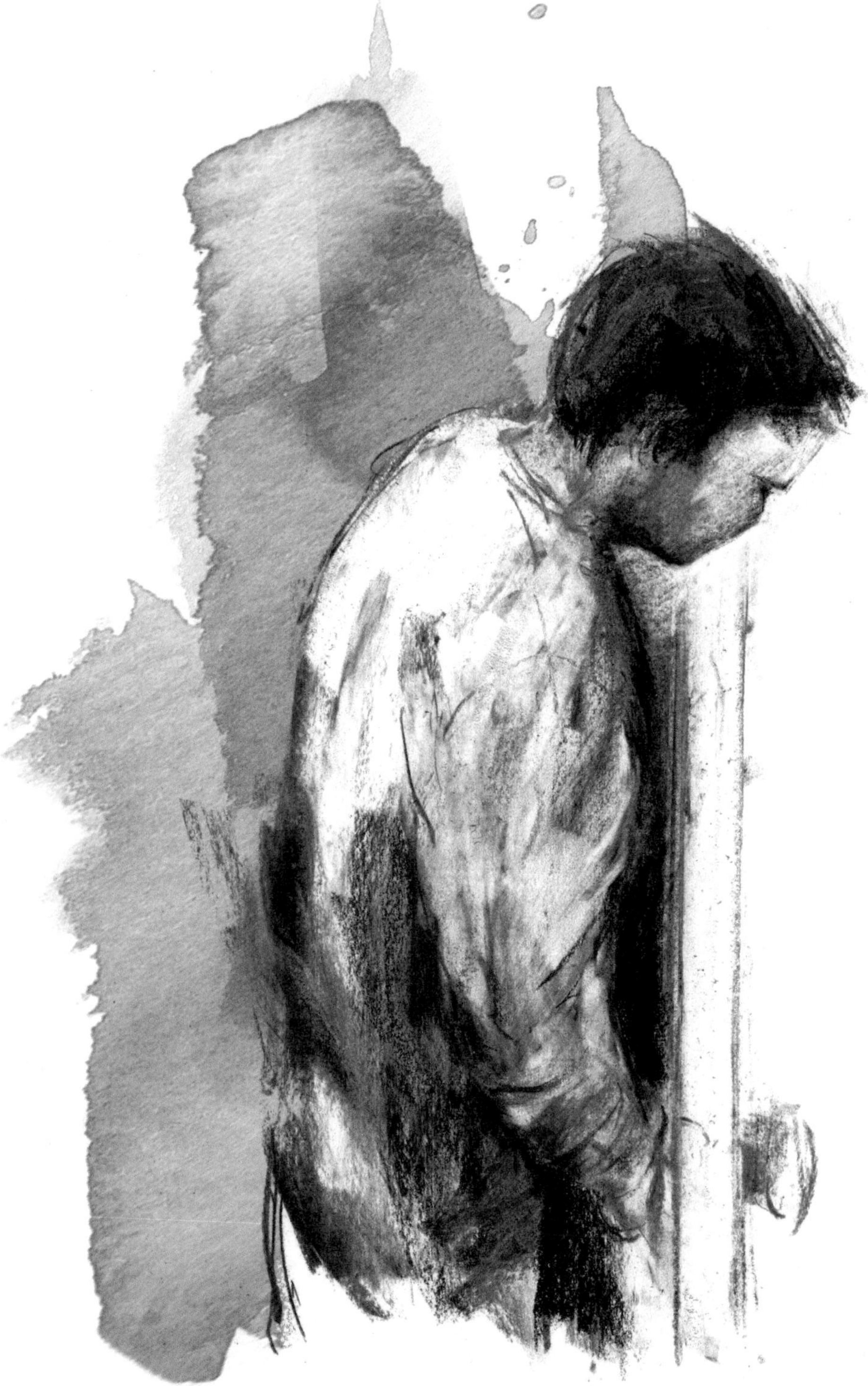

Who are *you*?

> *I don't know, but I found this.*
> *Can I come in?*

Do you want to?
This isn't a happy house.

> *It can be.*

But I don't know who *I* am.

> *Does that matter?*
> *We can find ourselves together.*
> *Everyone gets lost sometimes.*
> *But no one is everyone . . .*
> *you're just* you.

See ... happiness is easy.

Where did you come from?

From the light.

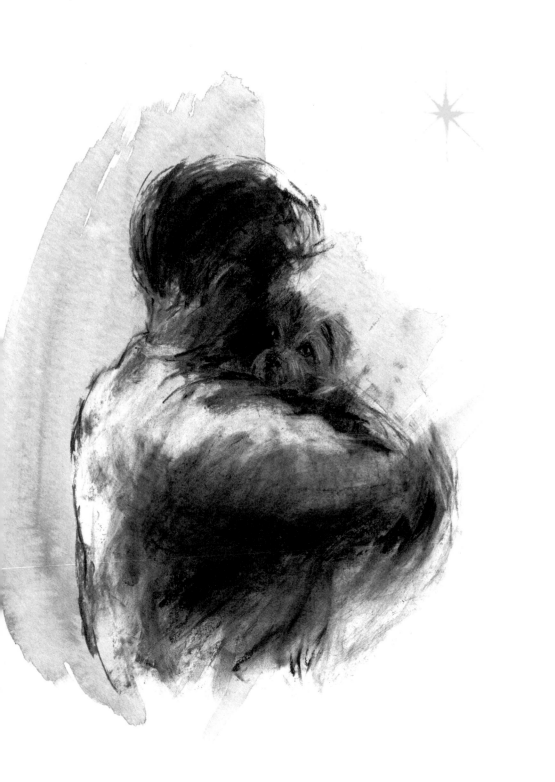

This is the light.
You can't touch it. You can't smell it.
You can't taste it. You can't hear it.
You can't see it.

I can feel when it's gone.

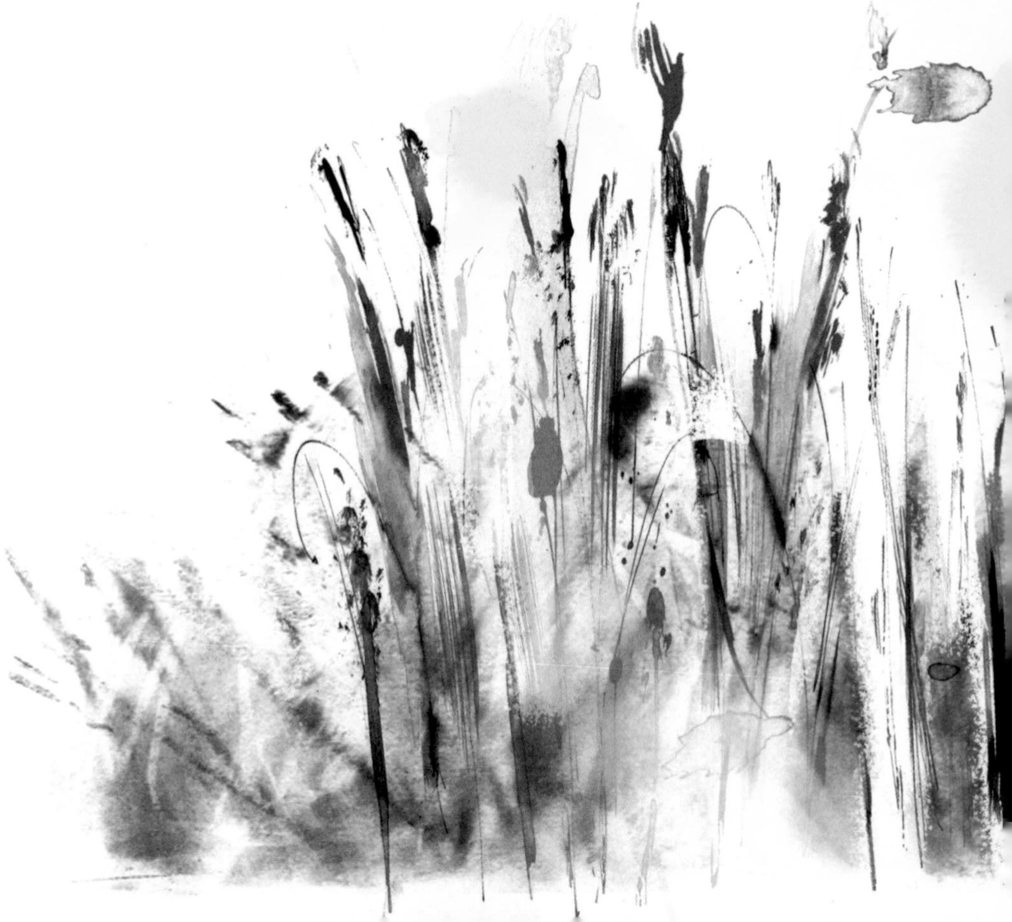

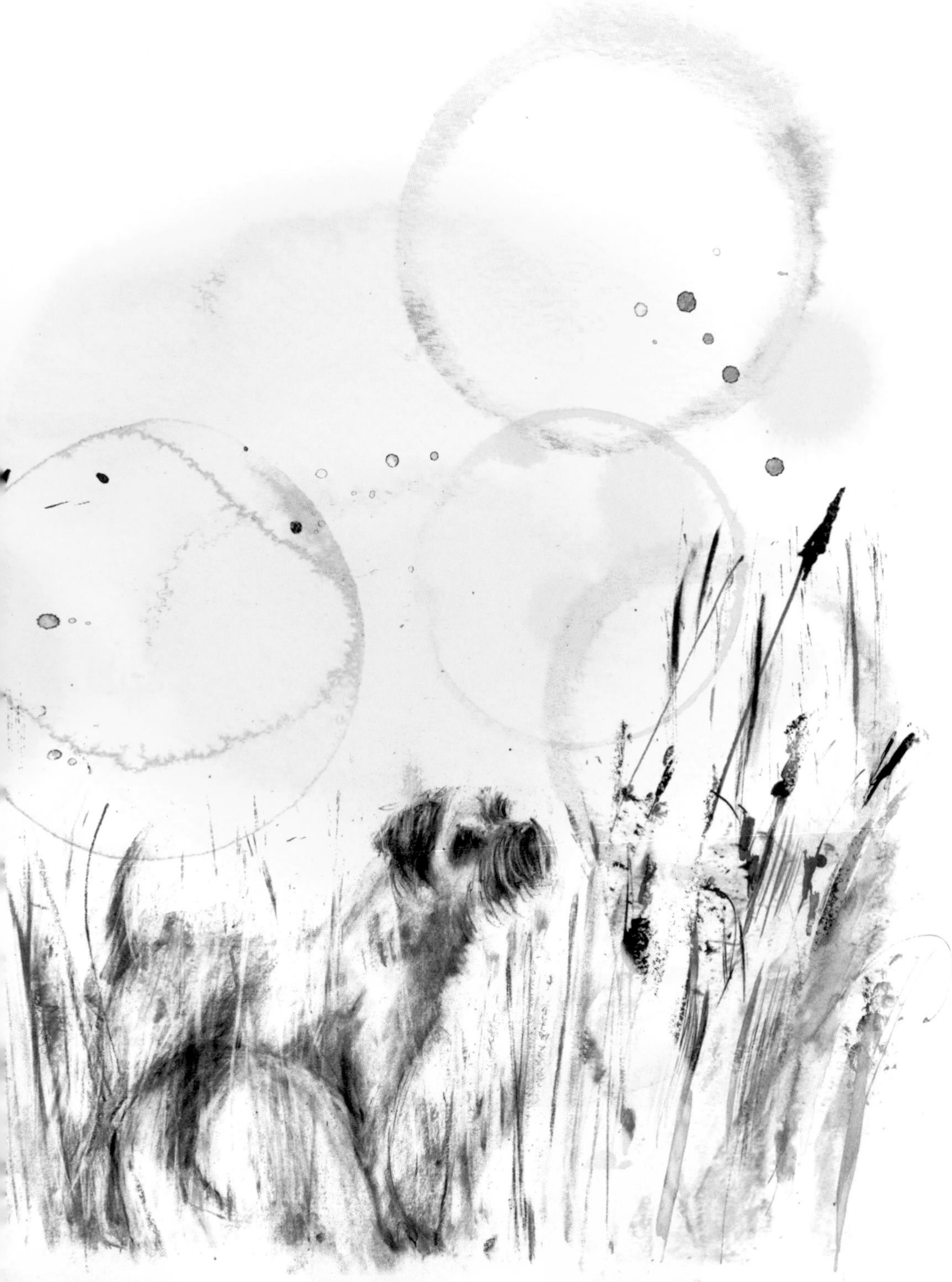

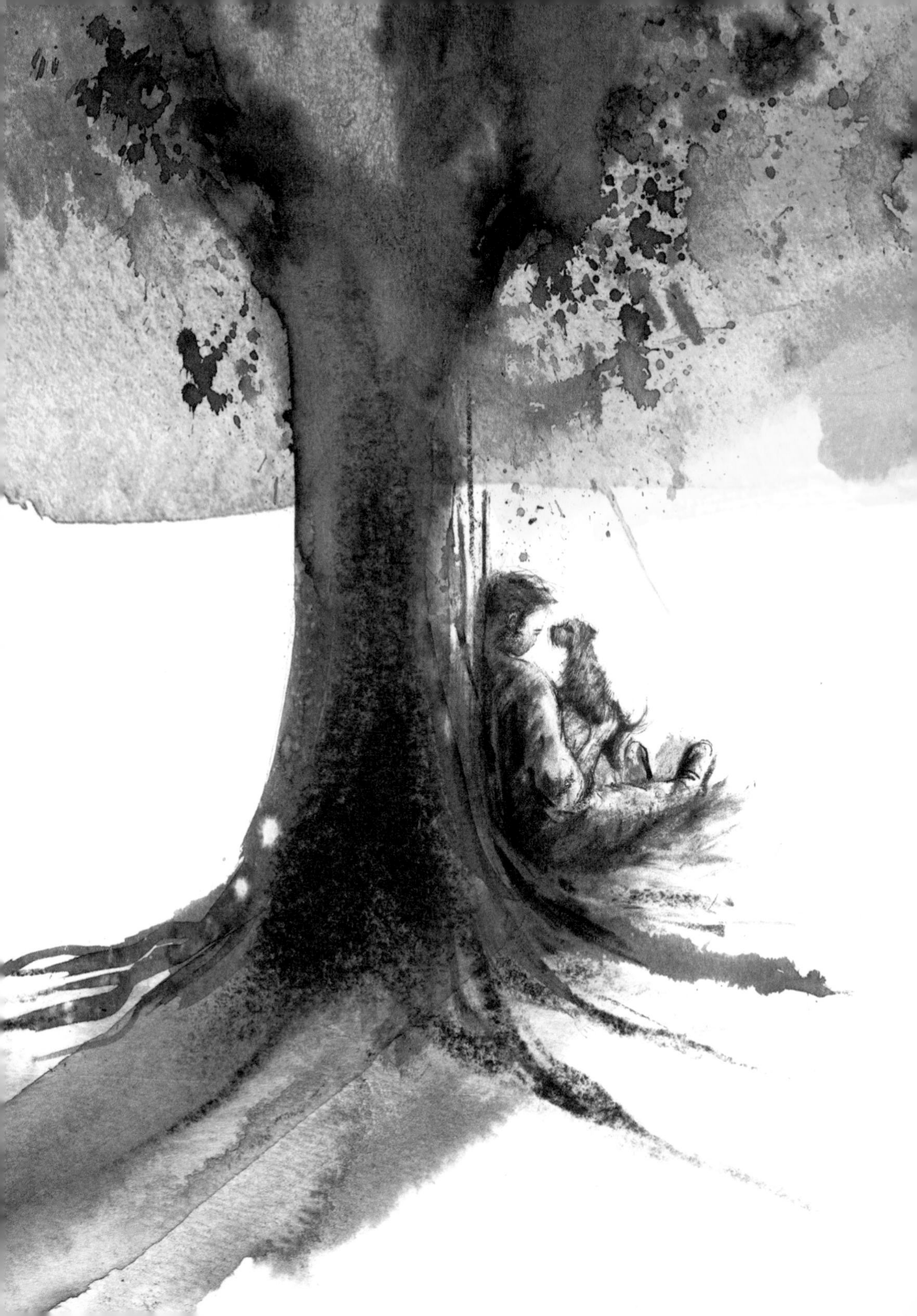

I'm afraid of the dark outside.

I'm afraid of the dark inside.

We can be a light for each other.
One of us can be strong
when the other is weak.

Tomorrow is a better day
because you're in it.

I just don't want to lose you.

That's silly.
You can't lose what's always there.

What's that?

Love.

Am I yours now?

Well, people call
me your 'owner'.

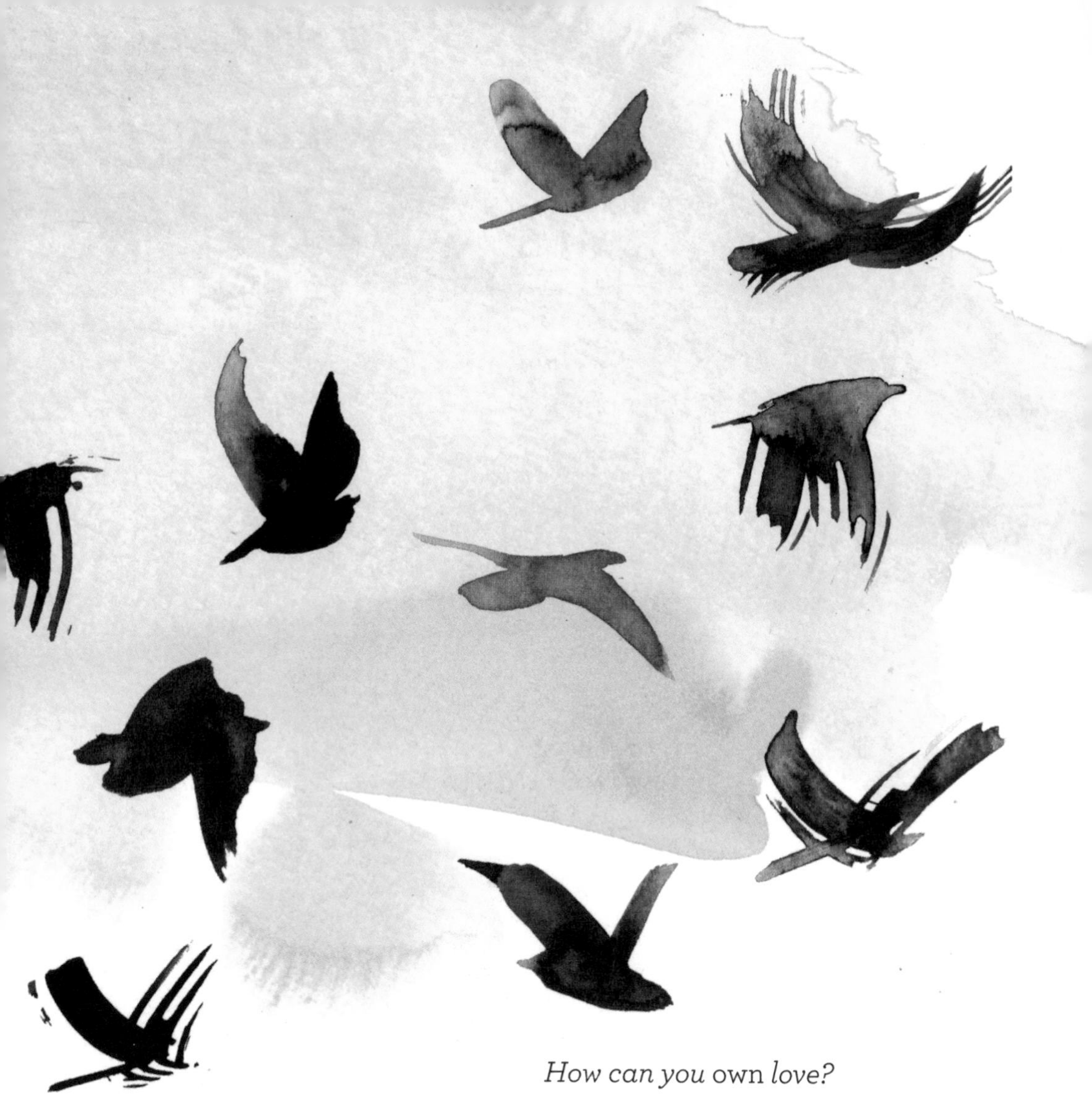

How can you own love?

You can't, I suppose.

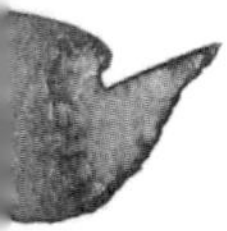

Can we just be ourselves and
still be each other's?

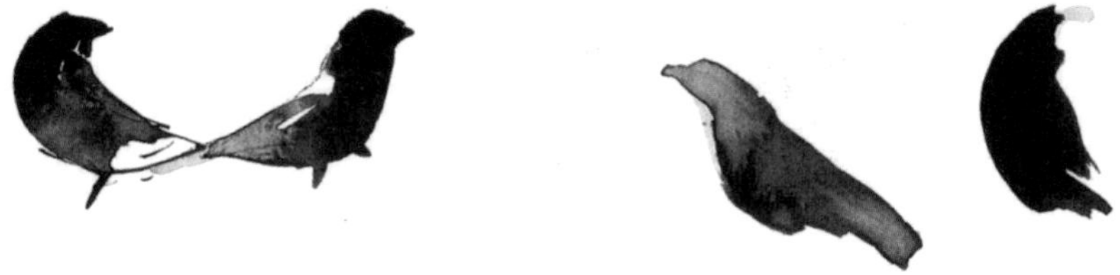

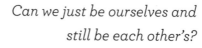

Slow down! That's a hedge.
There are thorns!

It's a hole . . .

You want to run and jump
into everything.

Is that bad?

There are dangers.
You might get hurt.

But I might discover
something wonderful.

Be careful – everything's so fast . . .

That was cruel.

What is cruel?

The opposite of kind.

Why are people cruel?
Is it harder to be kind?

No, it's easier to be kind.

So why isn't everyone
kind, then?

Look at all the different
kinds of people, Keira.

Is that my name ... Keira?

Well, only if you like it?
It meant *little dark one* in olden times ...
but *glittery* or *shiny* in Japanese.
In Russian it means *one that people
look up to.*

*Why do people look different and
speak different languages?*

Because they come from different places.
We all see the world in our own way.
But ... on the inside ... we are all the same.
My mammy taught me that.

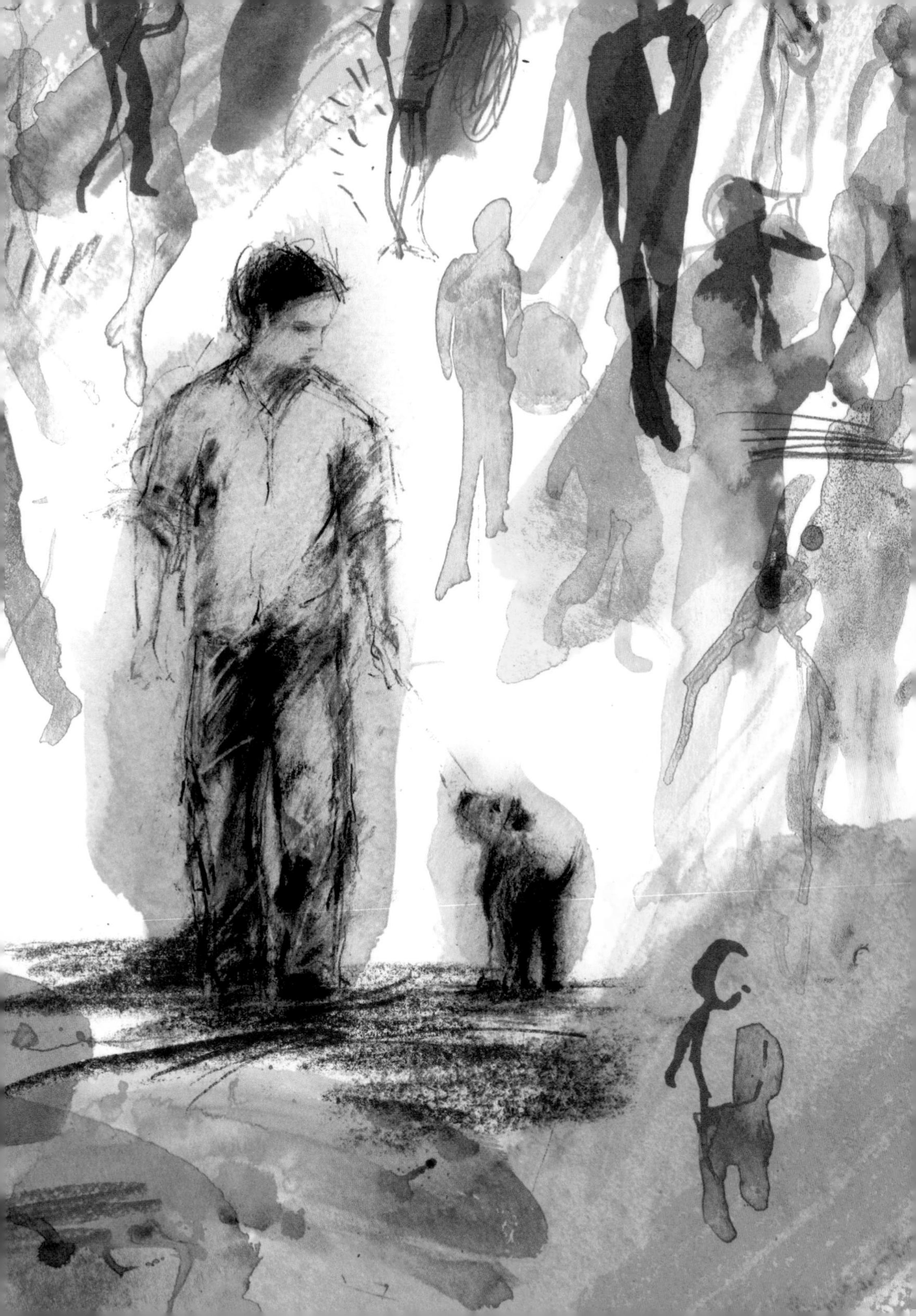

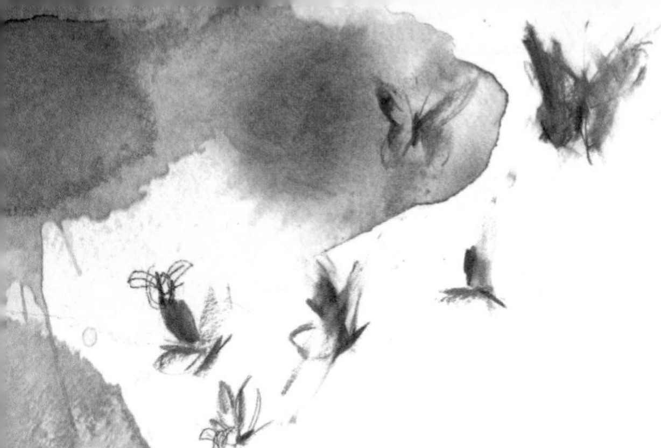

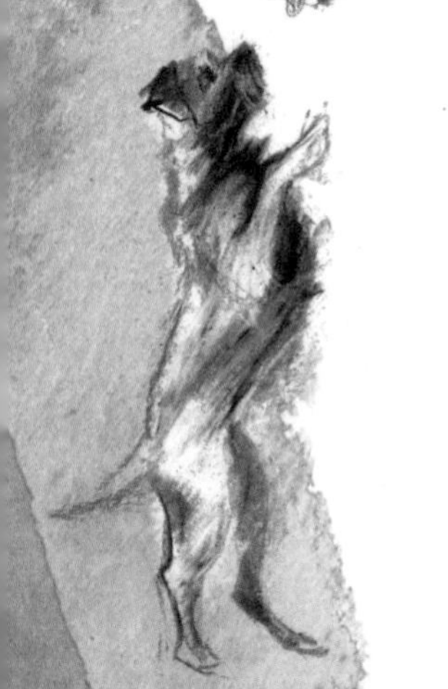

Do they *feel they're different?*

I don't know. Probably not.
I think they realise their 'sameness'
more than humans do.

I want to fly, climb, run and swim.

Dreams give you something
great to try for,
but you might fail along the way.

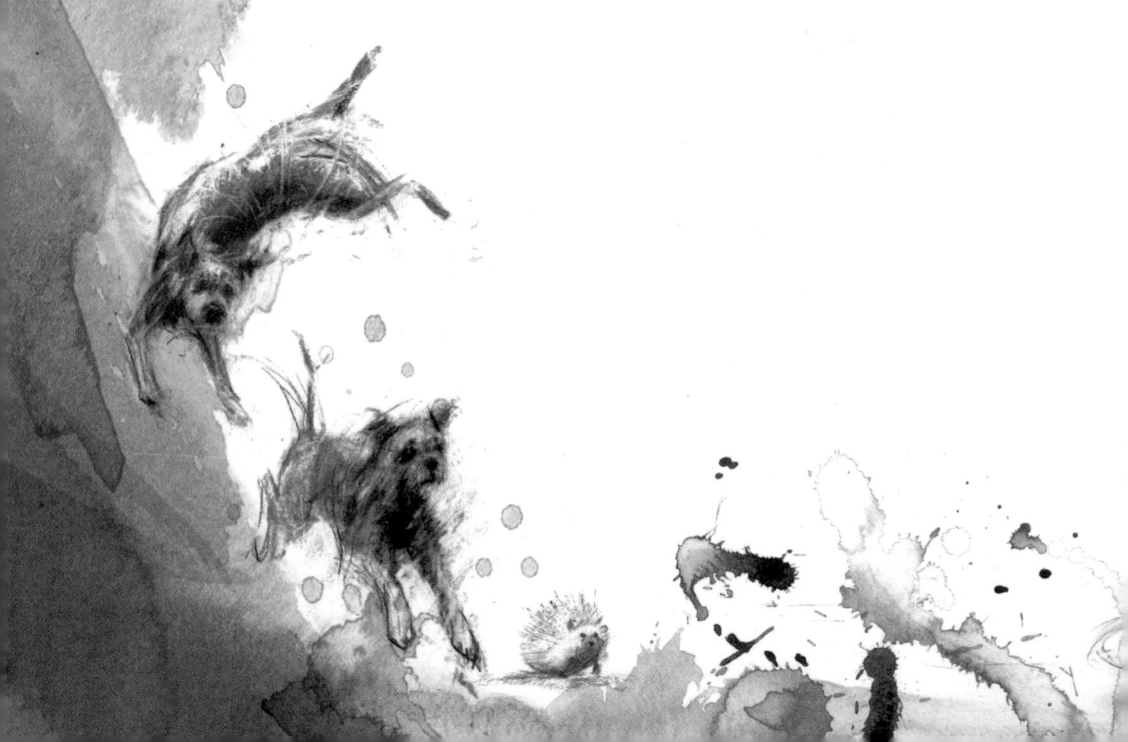

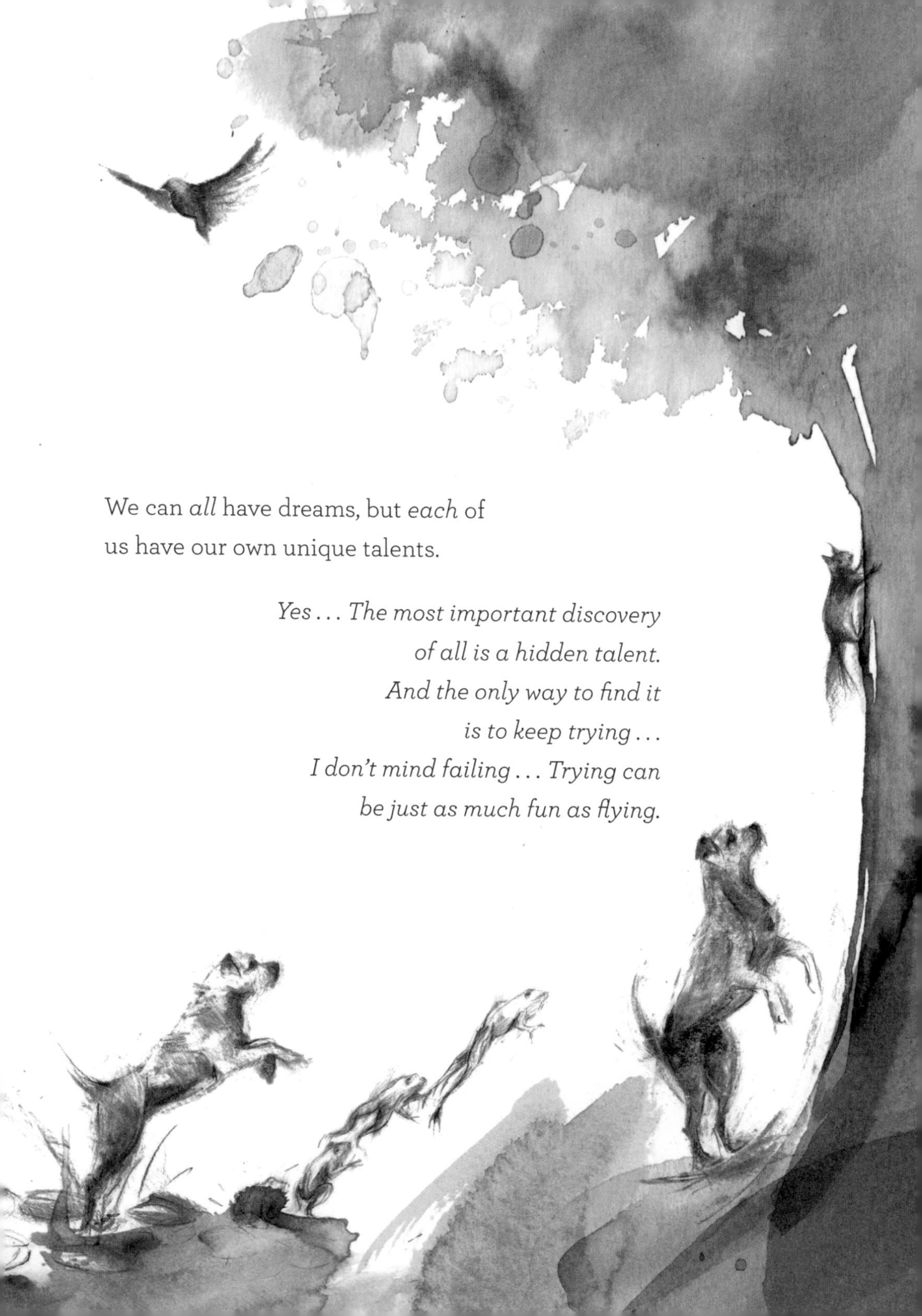

We can *all* have dreams, but *each* of
us have our own unique talents.

Yes . . . The most important discovery
of all is a hidden talent.
And the only way to find it
is to keep trying . . .
I don't mind failing . . . Trying can
be just as much fun as flying.

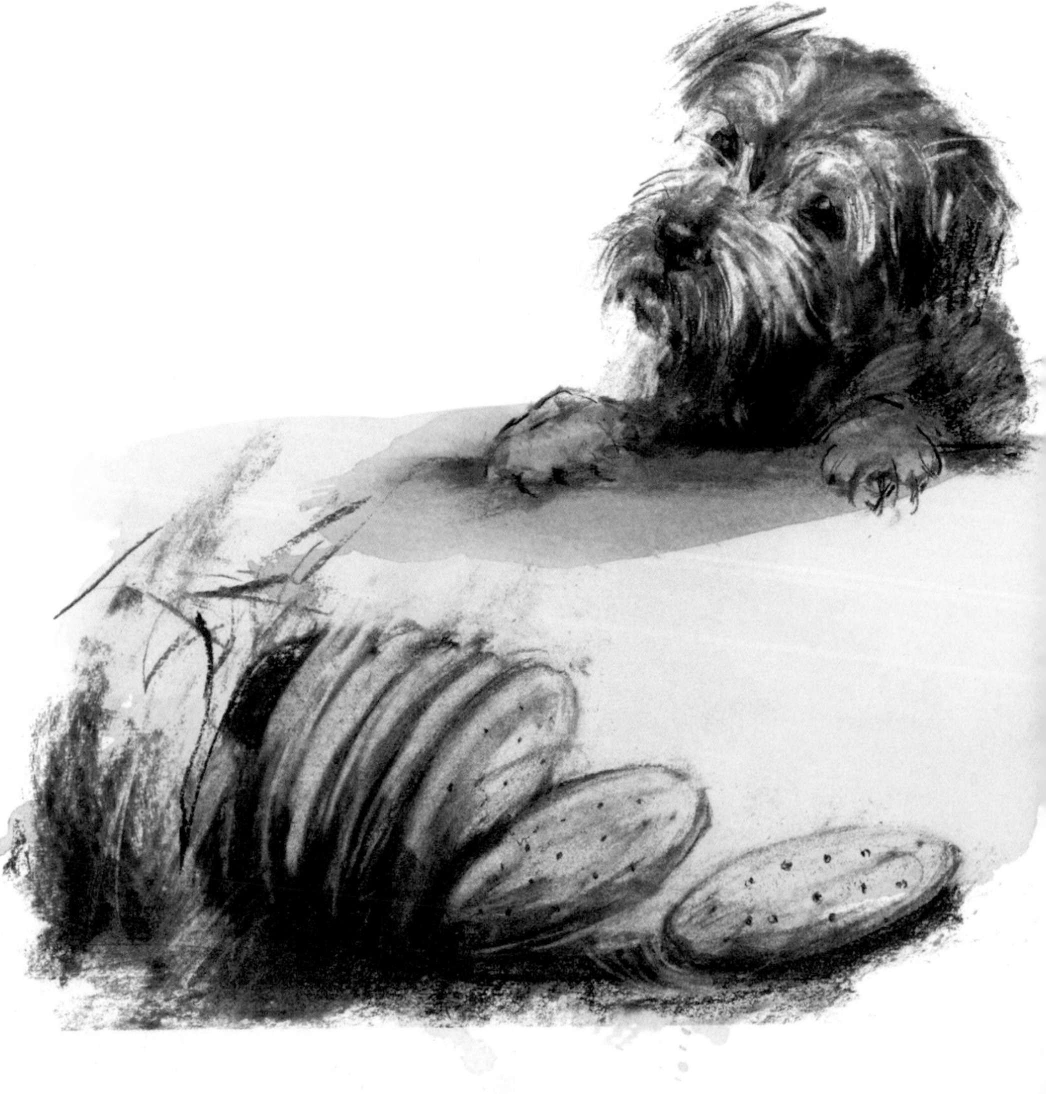

We can be different and still
be the same. We don't need
to agree on everything.

As long as we agree on the most
important thing.

Biscuits?

Love.

You're like my shadow, Keira.
I've got to work.

It's in your shadow that I shine the most.

You're always going to work.

It's who I am, Keira.

*Who are you without these
clothes, when you're with me?*

I don't know.
I don't feel good enough without
these clothes.

But you feel good enough
when you're with me?

Yes.

Well then . . . you're good enough
with no clothes at all.

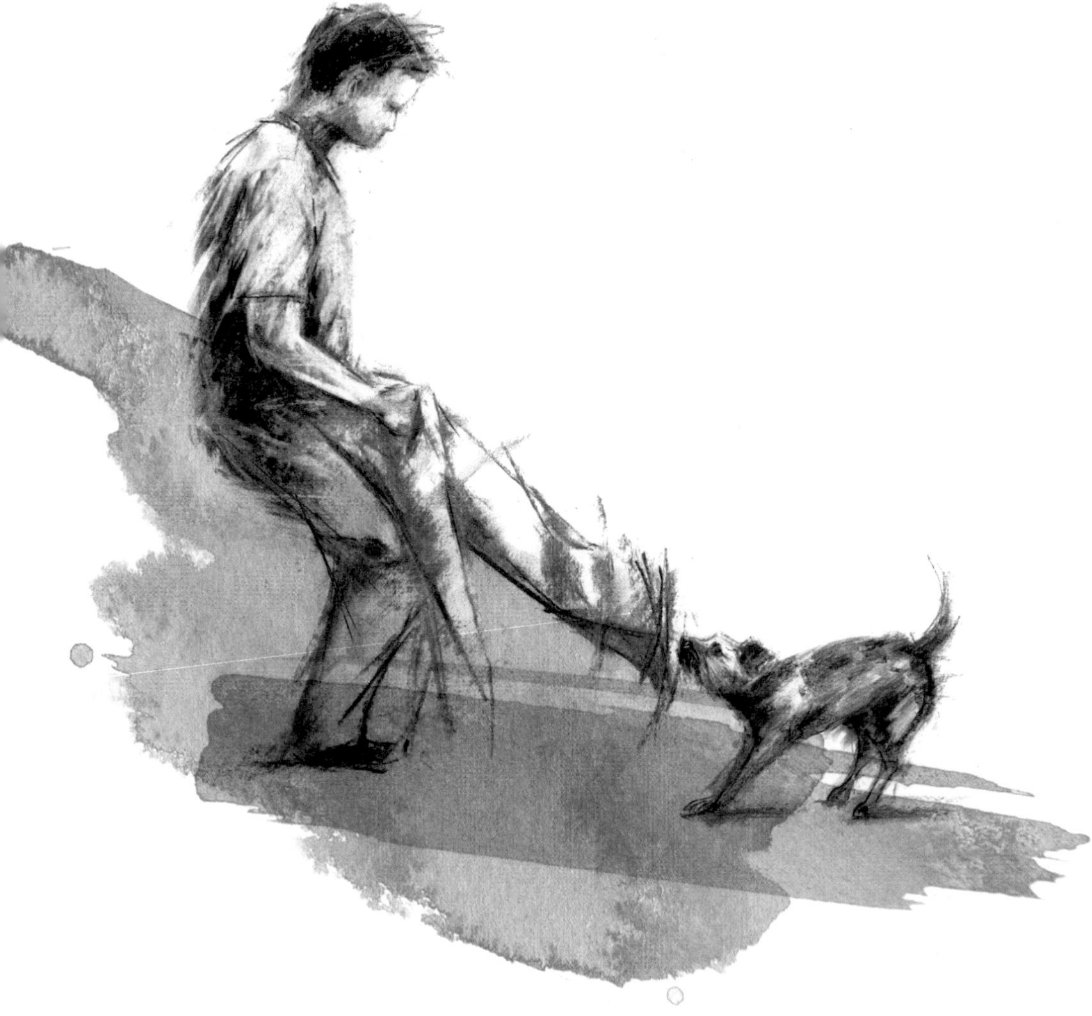

Don't worry. There's enough love
inside me for all of the animals.
Animals love me back, but
sometimes people don't.

Why?

Because they don't understand.
They judge.

But you're not who people think you are.
You're who you think you are.
And I love who you are.

You are doing the right thing.
That matters.

And you matter more than anything
else in the whole world to me.

That's where we operate
on the animals, Keira.

I hope I never end up in there.
It must be strange, to have
a life in your hands.

It feels like love.

What are those?

They're the tools I use
to fix the animals.

They look sharp. Does it hurt?

No. My patients are asleep.
I take away their pain.

*And what about when you are in
pain? Are there tools for that?*

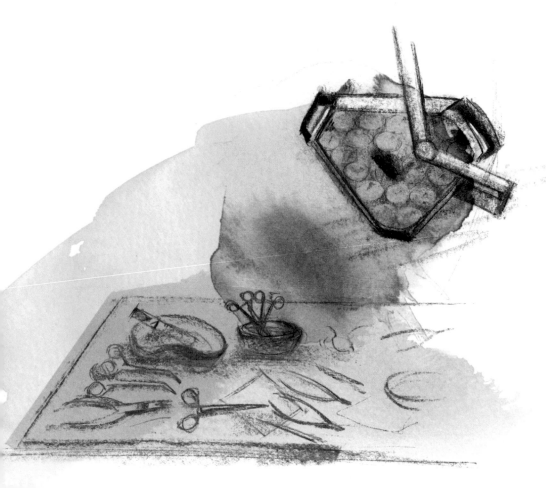

Do you fix all of them?

No, sometimes I fail.
Sometimes the pain is too hard
to fix, too hard to reach.

What happens then?

They might die.

So that's the end?

Well . . . no. Come outside.

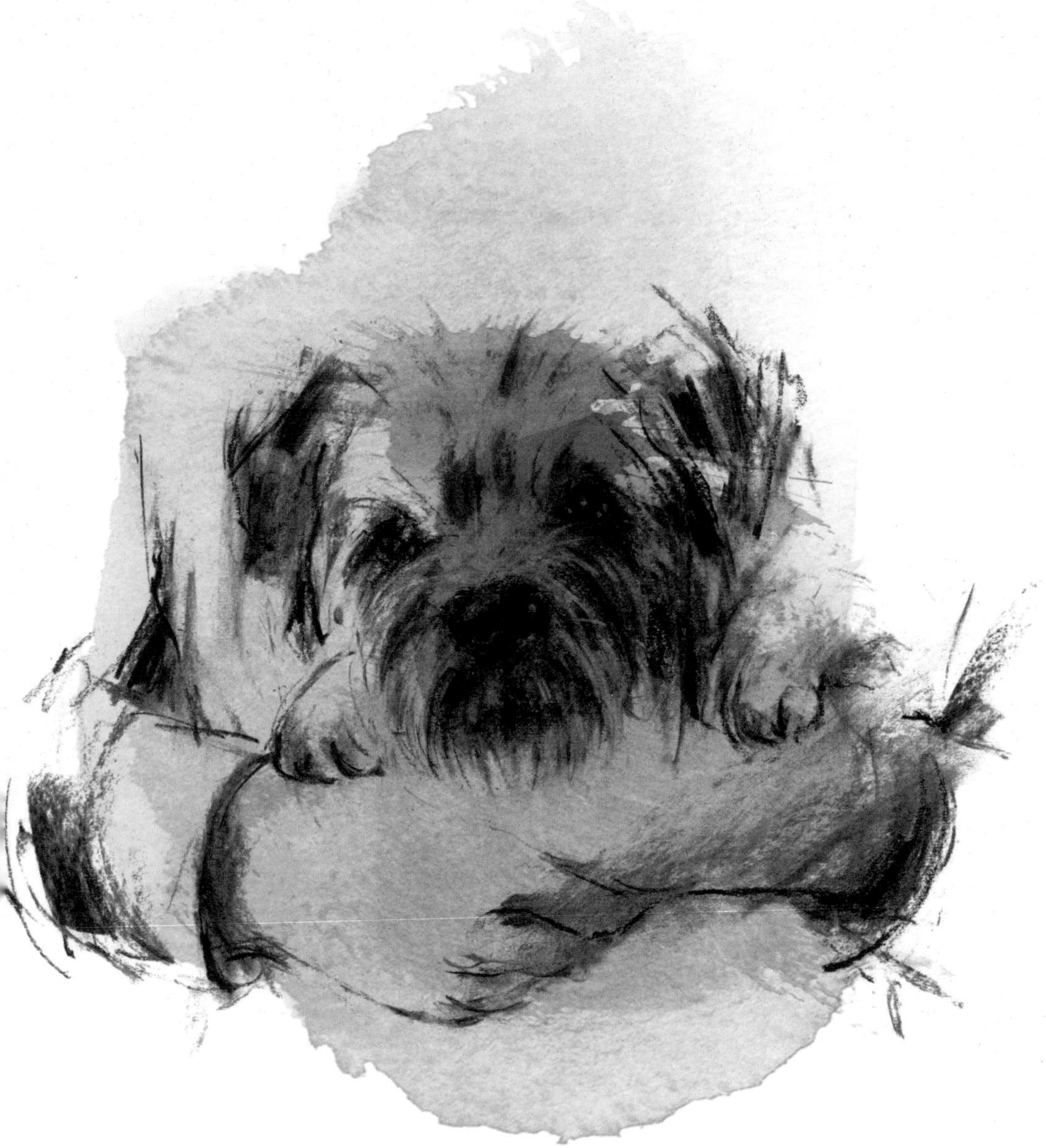

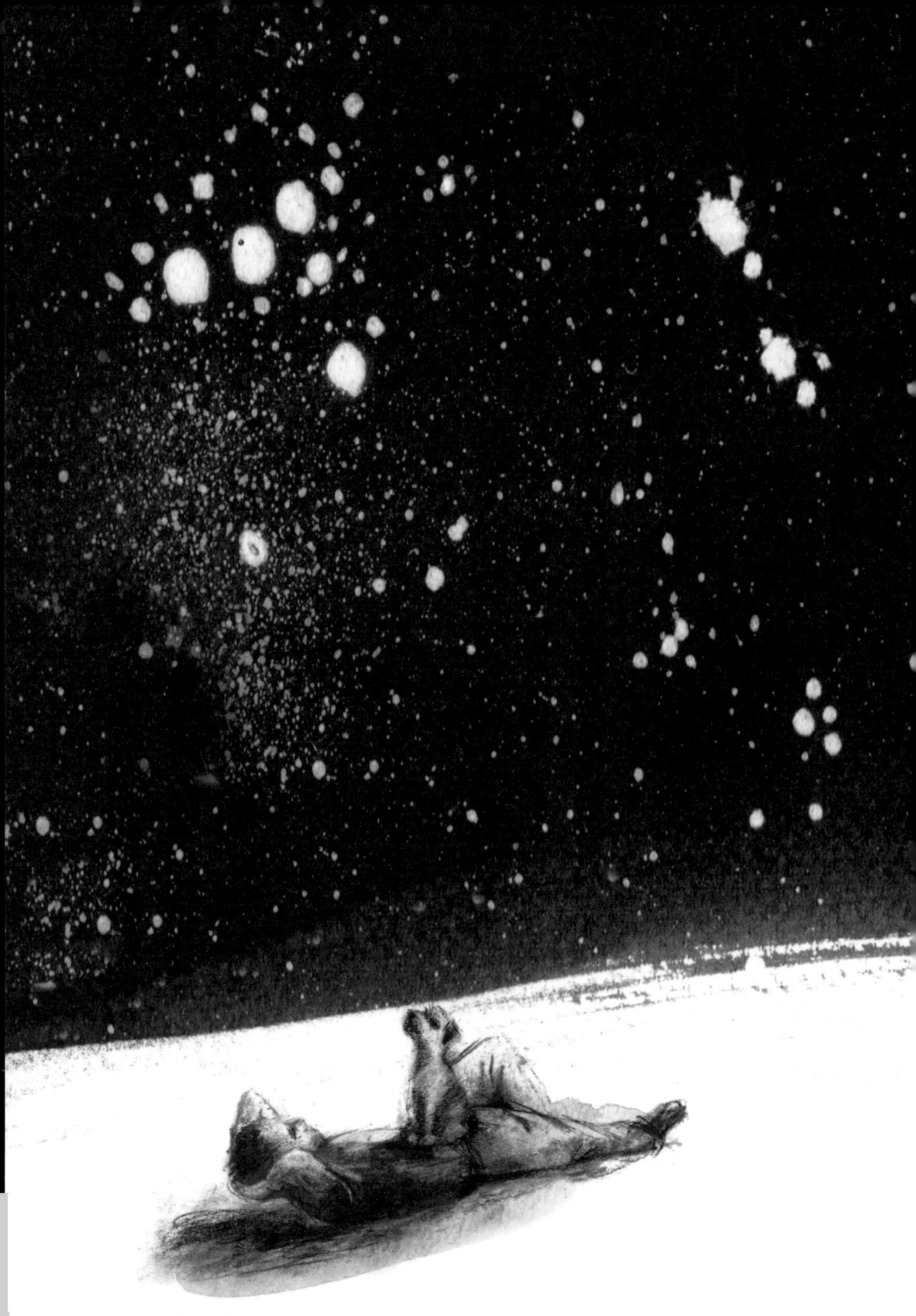

You see the stars?

They're beautiful.

That's where we come from.

Did we fall?

Sort of. We're all stardust,
Keira. We always will be.

And our light lives on?

Yes. But you can only *feel* it,
like the sun.

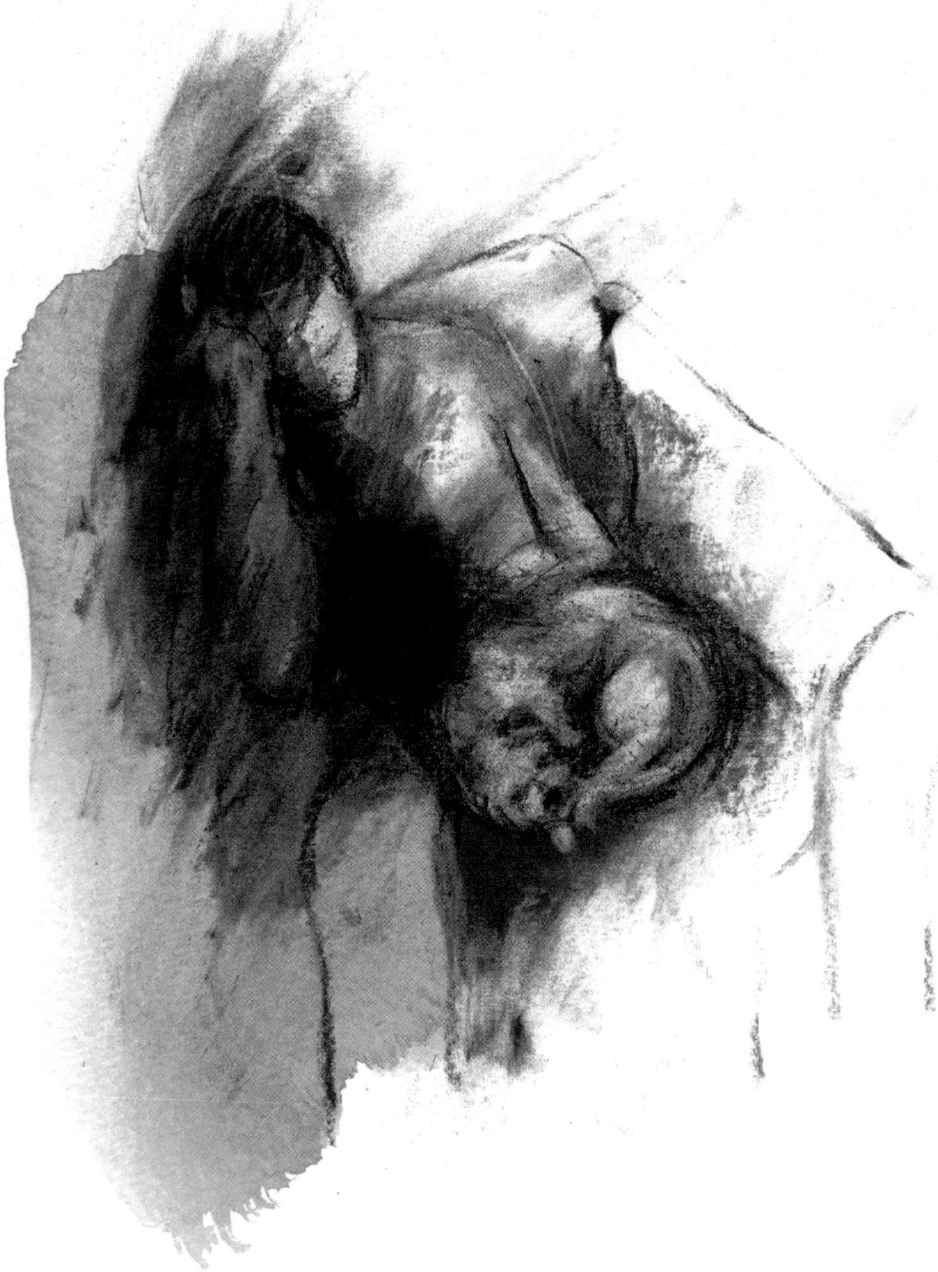

It didn't work out today.

> *But you did your very best.*
> *I can taste it in your tears.*

You don't judge me, Keira.

> *I wouldn't know how.*

Keira, it's raining!

You have to stand in the rain
to enjoy when it stops.

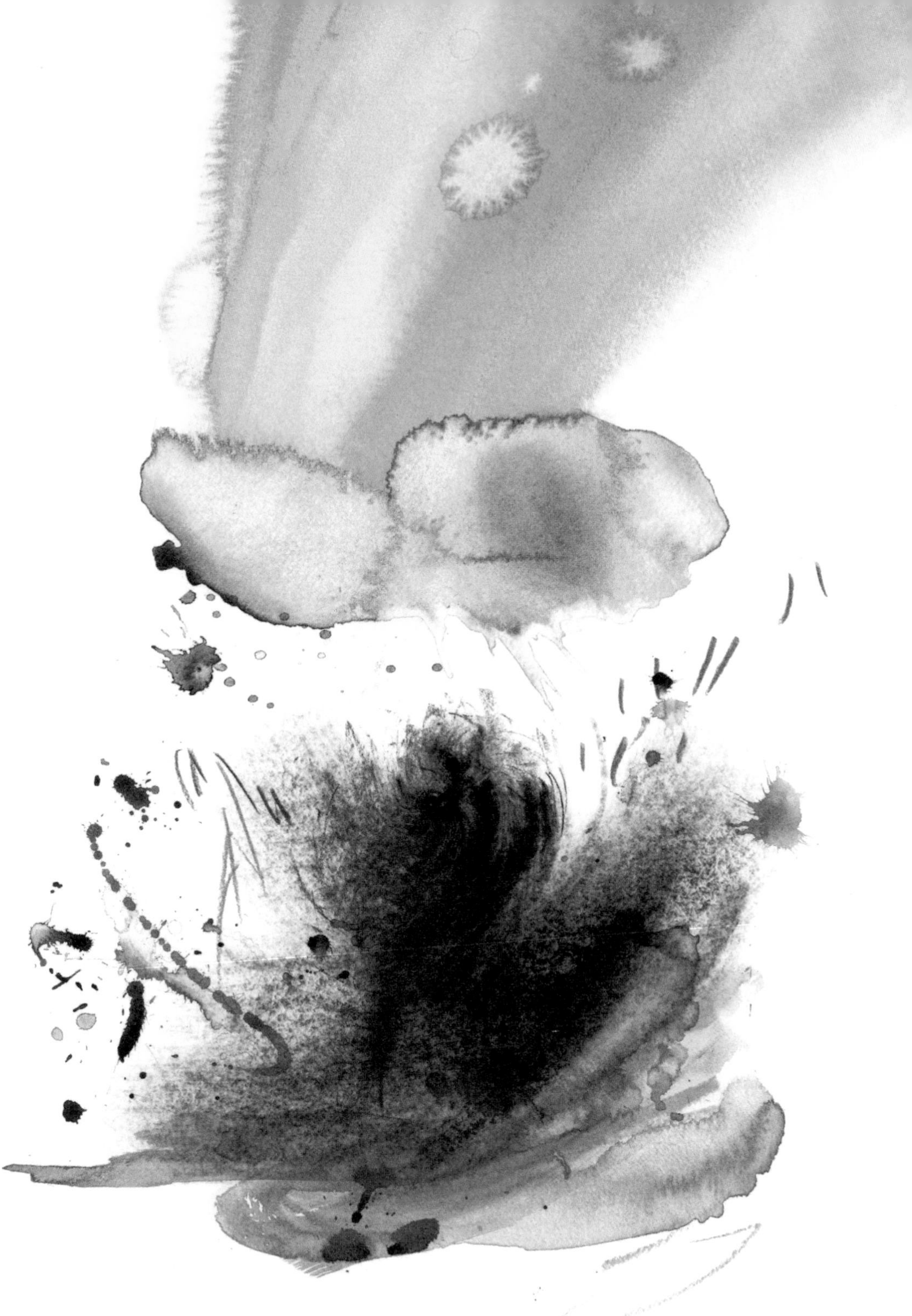

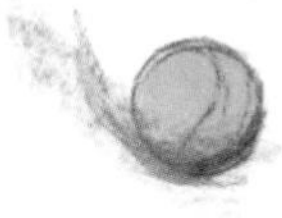

Throw it! Throw it!

It's a dirty, soggy tennis ball.

It's a world of endless possibility. Throw it!

I'm sorry. I threw it too far.

Don't worry. If it's meant for us, it will be there for us.

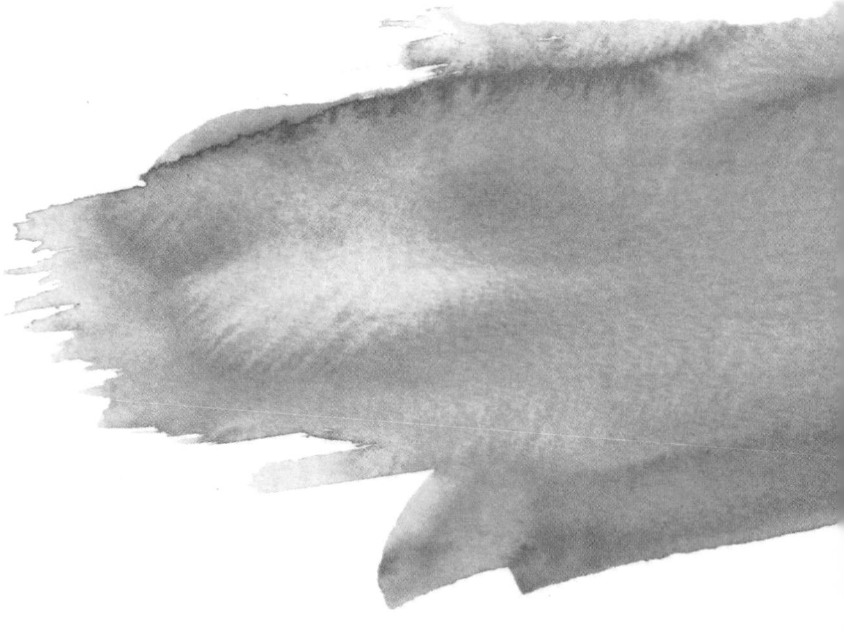

Don't people realise we need to look after each other – whatever our size or shape or colour?

I will always look after you . . . even when you smell.

Especially when you smell.

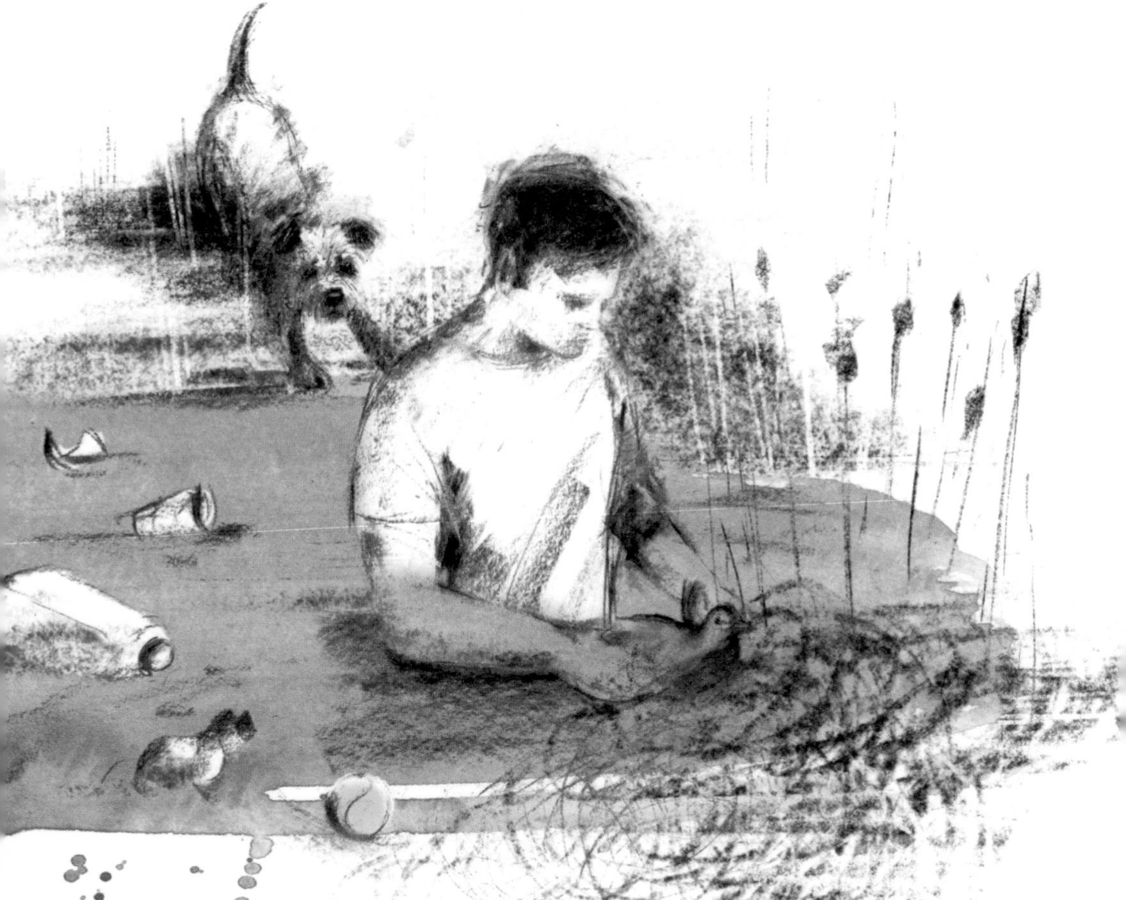

Being silent together was the best
conversation I ever had.

For thirteen years, your smile made
every day a good day.

For thirteen years, you taught me your
lessons in your own inimitable way.

For thirteen years, your unconditional
love lit up my world,
and I desperately wanted you to stay.

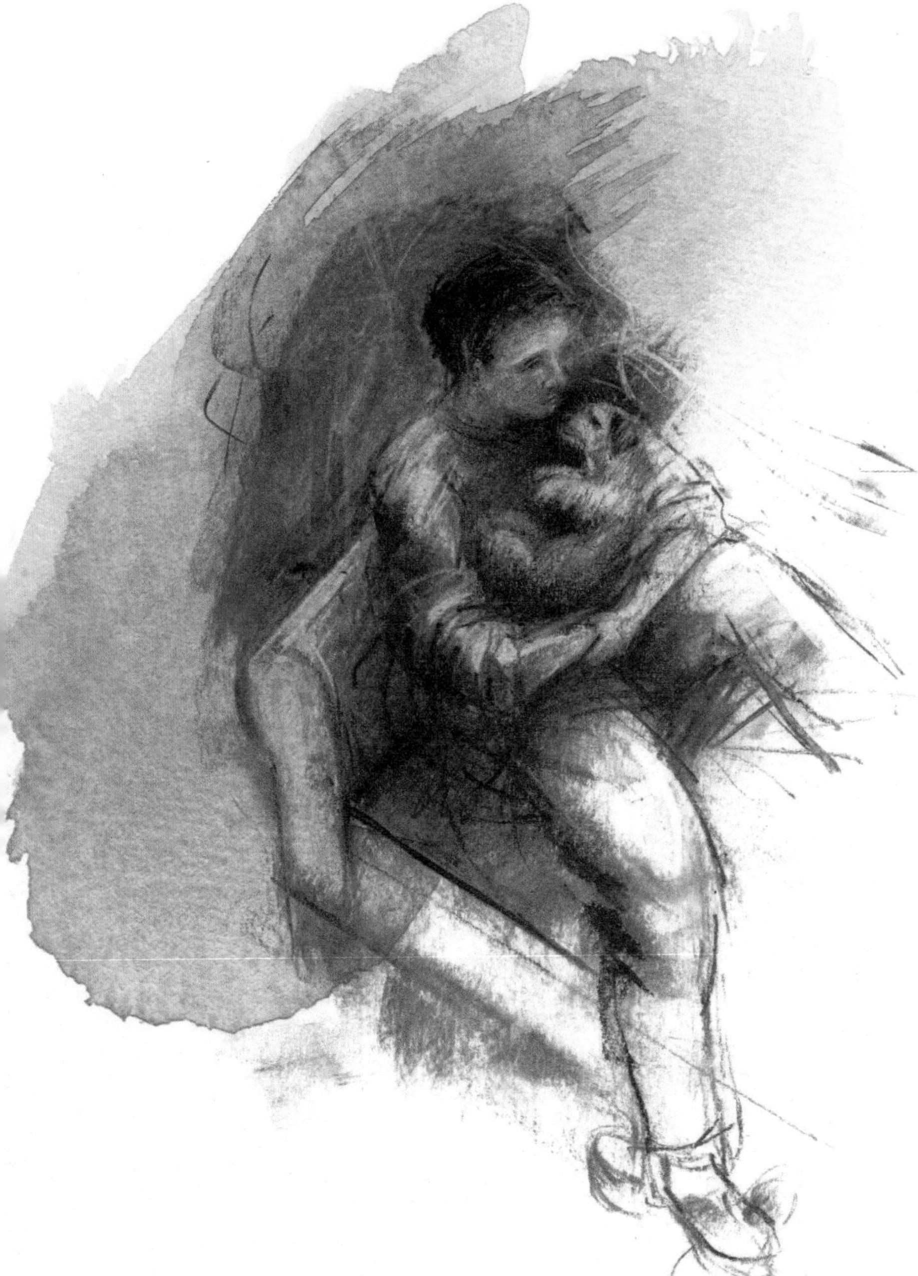

The night was starless.
I was lost in my thoughts.
I heard it too late.
I saw it too late.

I should have been walking at your side.
I should have been holding your paw.
But I wasn't.

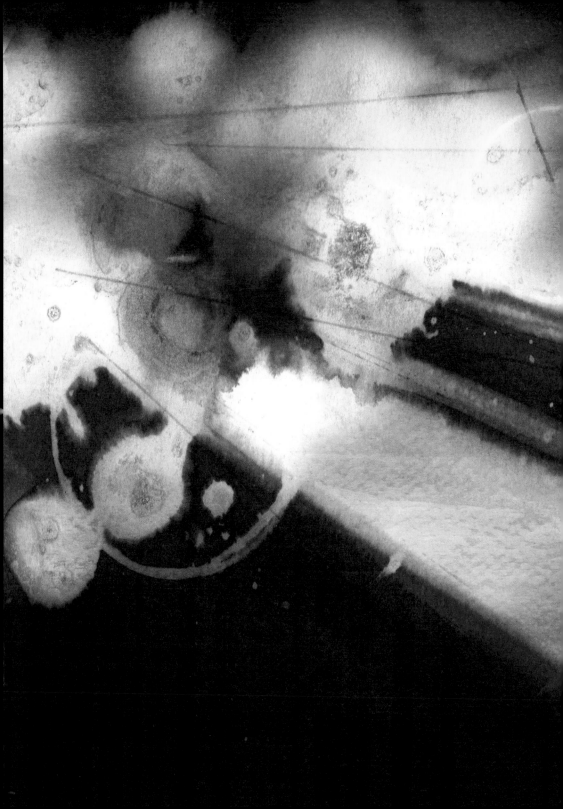

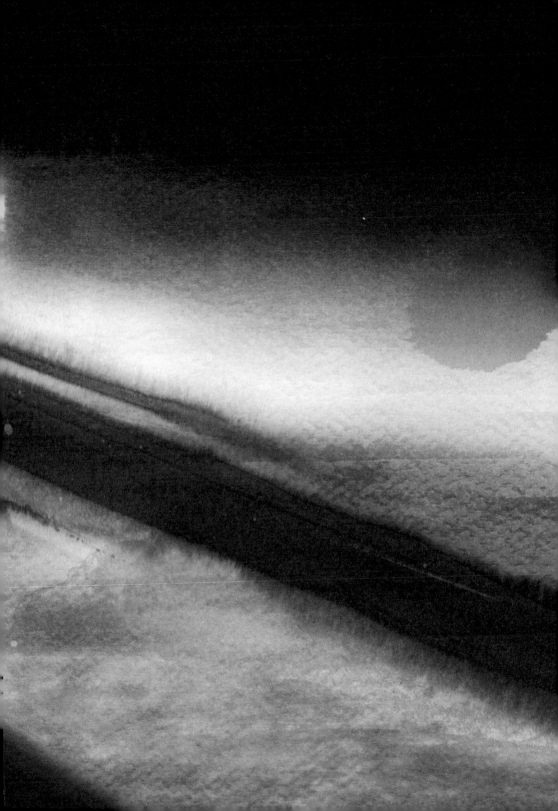

Great pain is the price of great love.
I spent my life wanting to be needed,
yet when you needed me most, I let you down.

I'm so afraid, Keira.

You taught me that it's not what happens,
but what we do next that defines who we are.

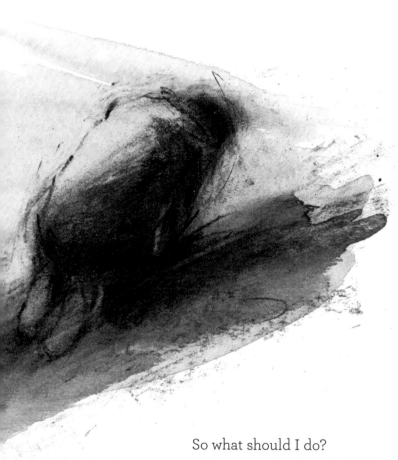

So what should I do?
How can I let you go back to the light
when I still see light in your eyes?

Am I holding on for me or for you?

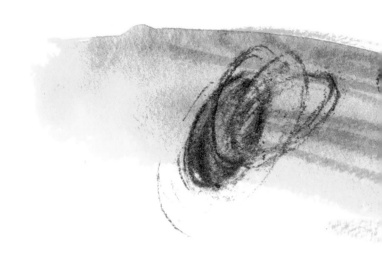

When I was broken
you bound me back together.
And now, all I know
is that I desperately want to fix you.
As you have fixed me.

Your fate is a spinning coin.
Heads or tails, success or failure, life or death.
Truth and love will always face each other,
never one without the other.

I watch it spin.

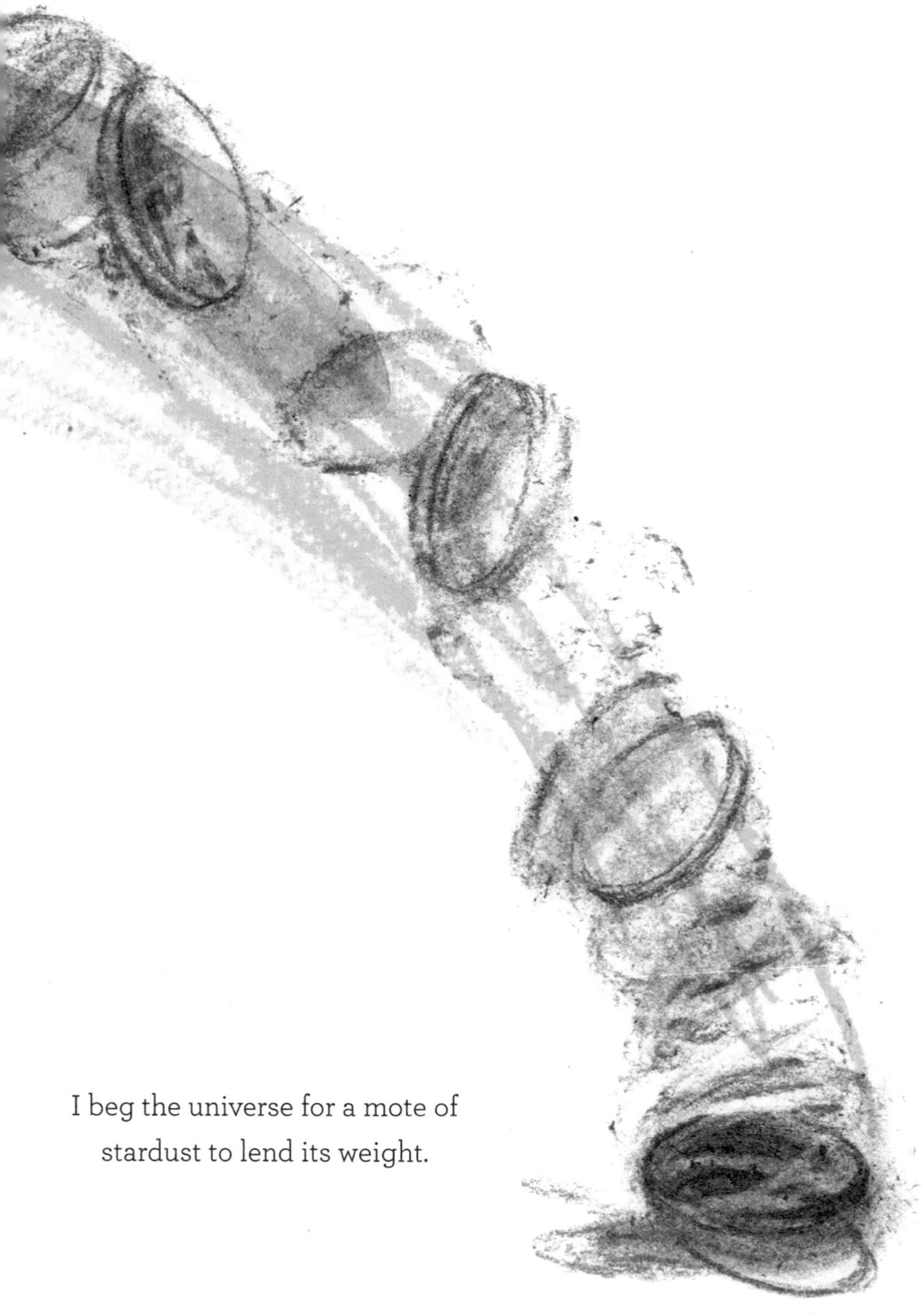

I beg the universe for a mote of
stardust to lend its weight.

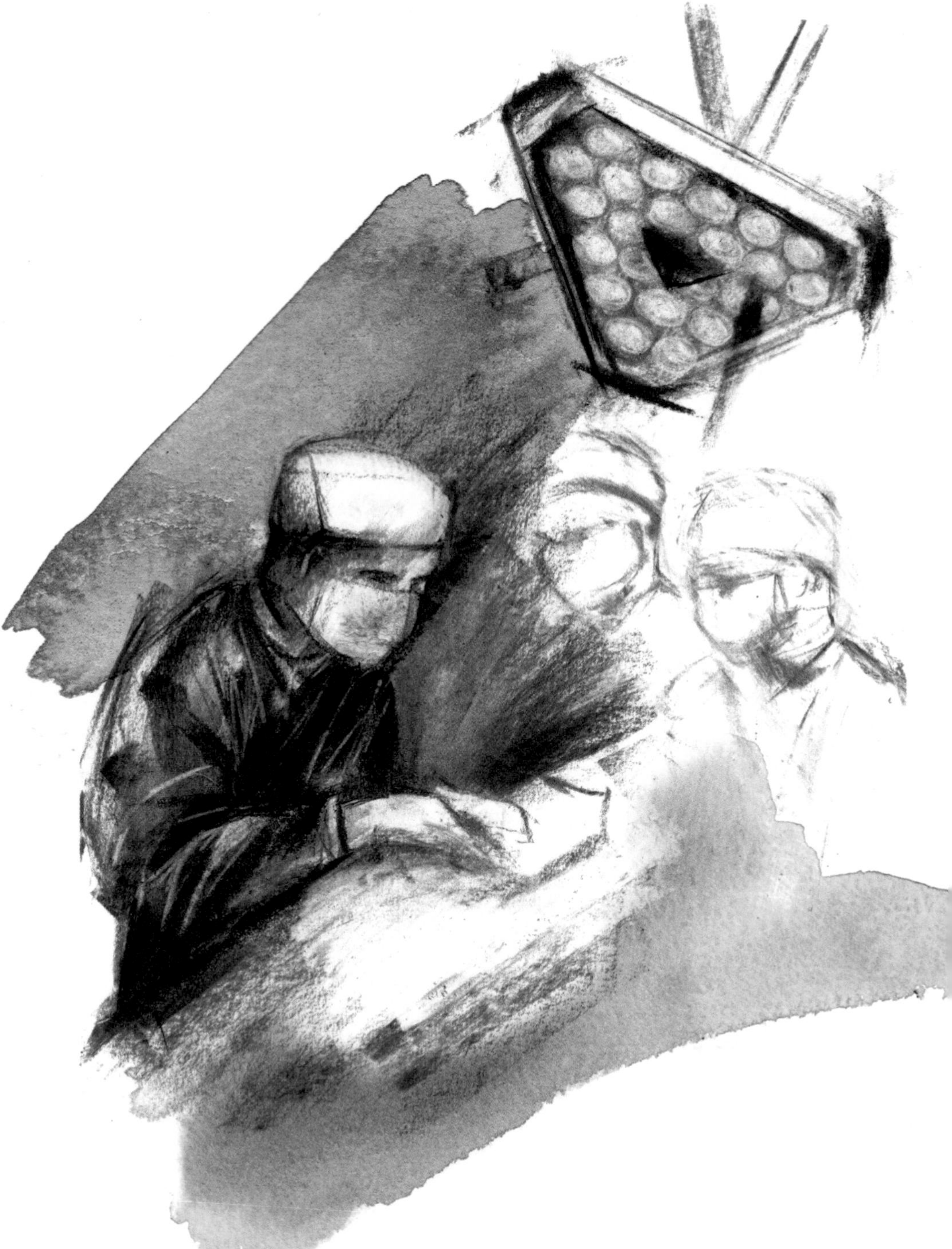

You were there when I didn't
even know I needed you.

You held my fear in your paws.
And now I hold your fear in mine.

The dark inside has come back.
I'm scared now.
I need to be enough . . .
for both of us.

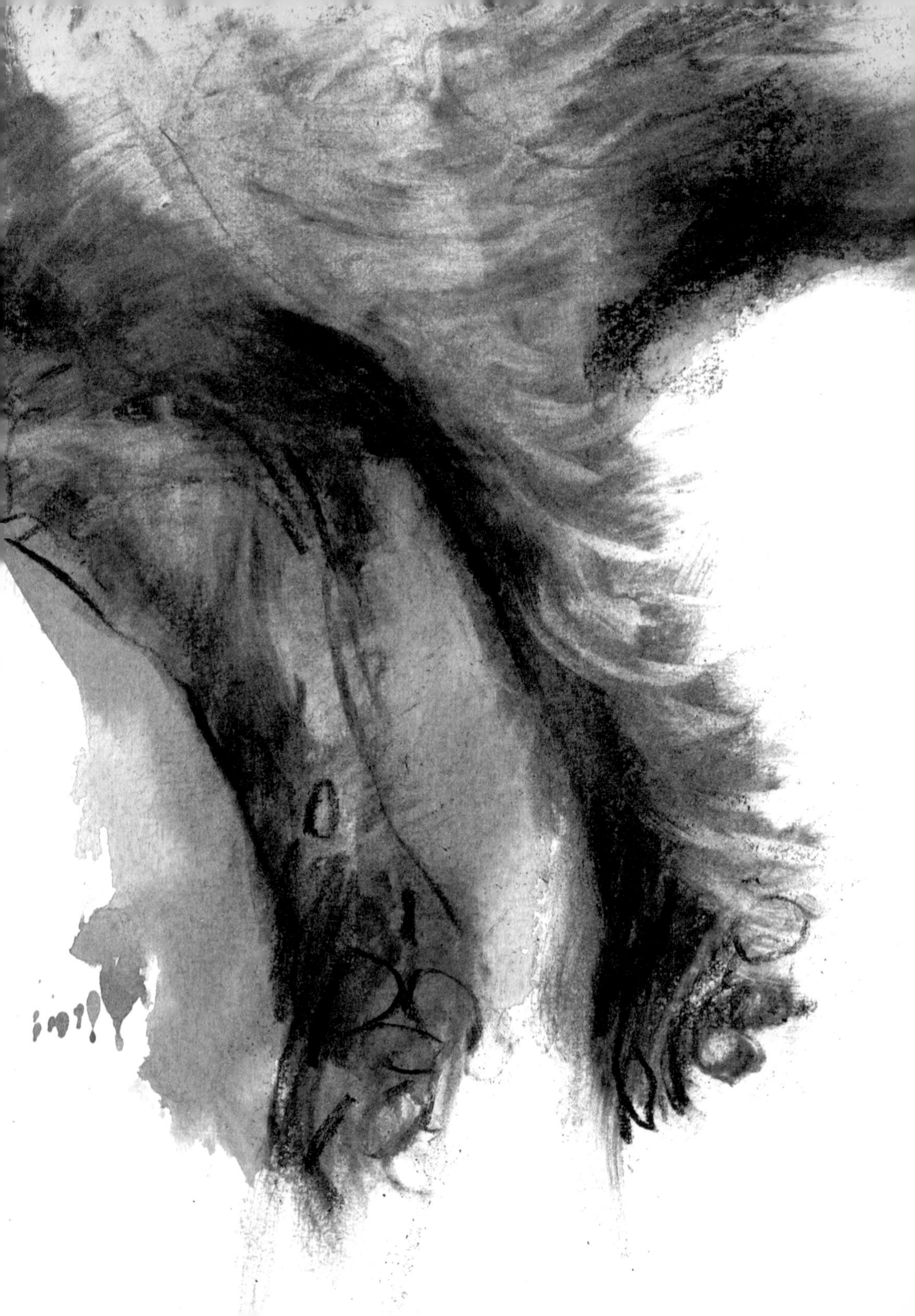

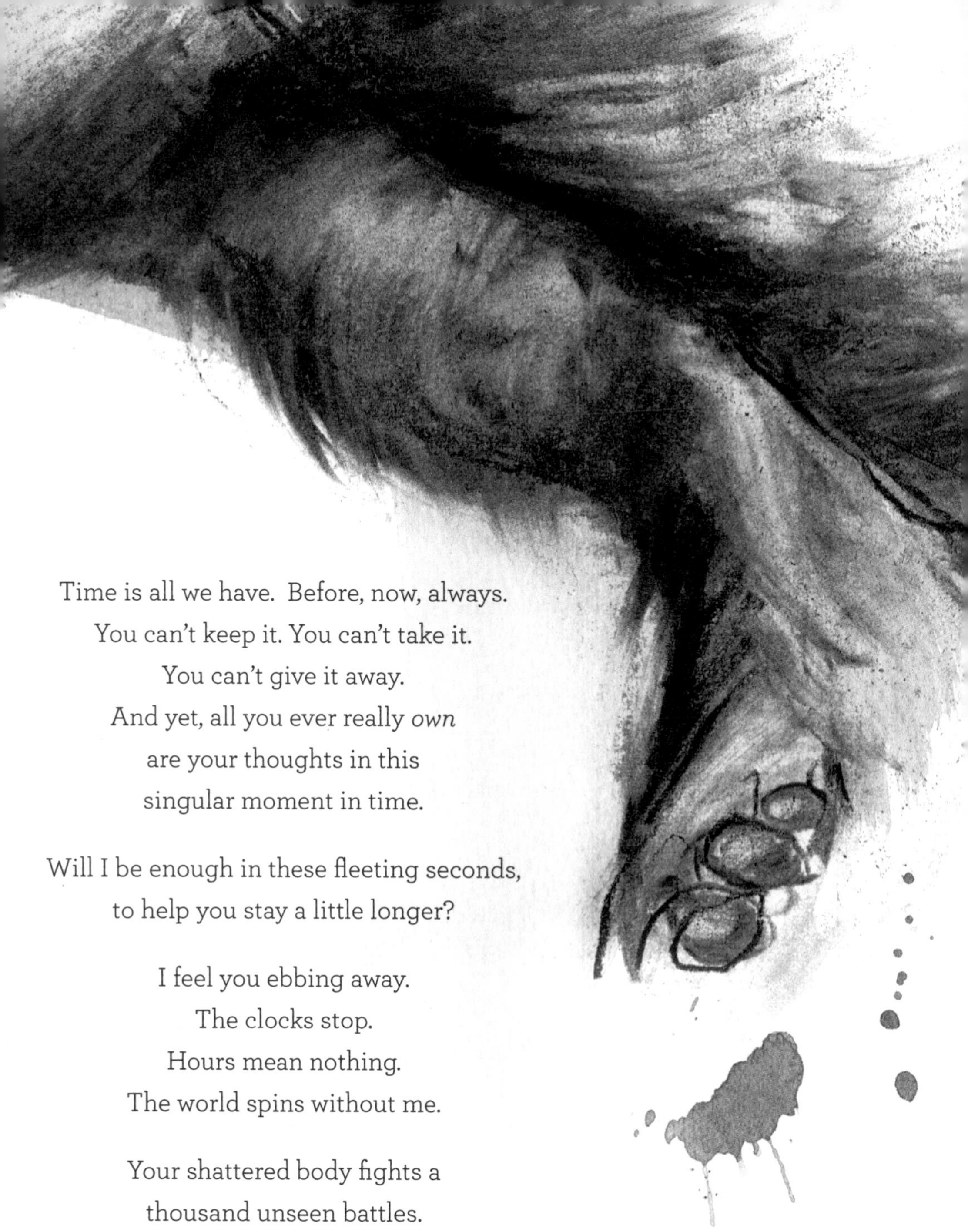

Time is all we have. Before, now, always.
You can't keep it. You can't take it.
You can't give it away.
And yet, all you ever really *own*
are your thoughts in this
singular moment in time.

Will I be enough in these fleeting seconds,
to help you stay a little longer?

I feel you ebbing away.
The clocks stop.
Hours mean nothing.
The world spins without me.

Your shattered body fights a
thousand unseen battles.
My shattered heart takes strength
from your courage.

Time smiled.
You came back piece by piece,
just as you had rebuilt me.
A nostril flaring to my scent.
An eye cracked to the light.
A tongue tasting a second chance.
A tail wagging, despite everything.

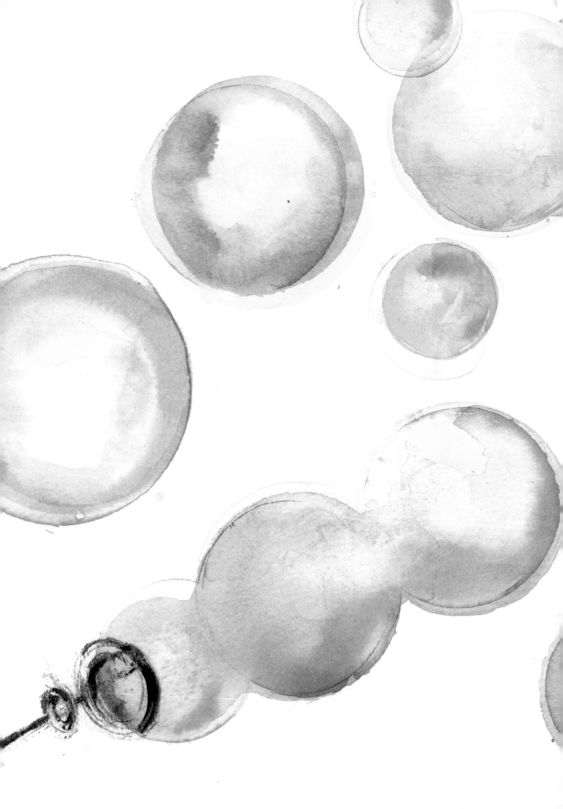

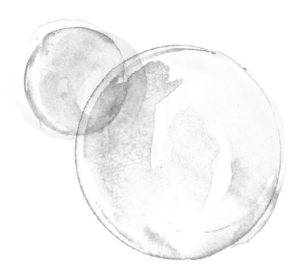

I will take your pain away, Keira.
Healing is like blowing a bubble –
blow too hard and it's gone.
'Take it easy . . . Not too fast!'

Hugging is half of healing.
If fear can drive us, rather than destroy us,
the thing we fear the most
can make us least afraid.

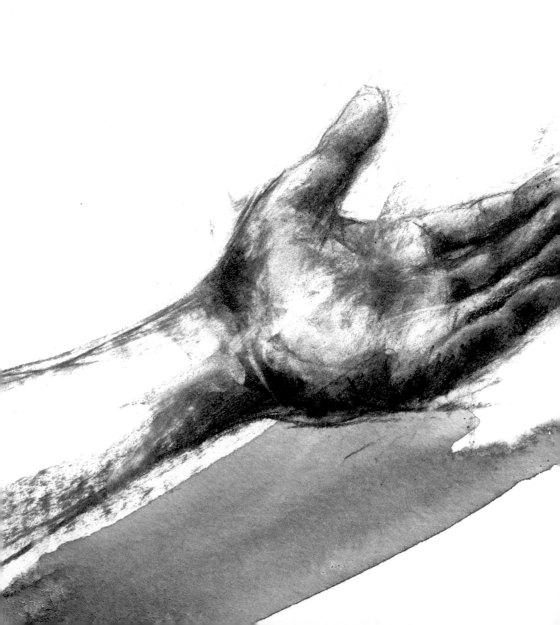

We can choose not to be afraid.
We can choose to really *live*.
But we cannot choose to stay alive.

Sometimes your hands hold back
the tide of time for a while,
before it trickles through the cracks.

You held your head high
and put your best paw forward every day.
Your tenacity shone like a beacon of hope.

You taught me so many things.
To balance the pride of triumph
with the humility of failure.
To weigh praise and criticism with an equal hand.
Not to seek credit for talents and blessings
gifted to me by the universe.
But instead to be always curious and grateful . . .
and just welcome them in.
To pay less heed to the voice in my head
and more to the voice in my heart.
And to know that your paw in my hand
is all the truth I will ever need.

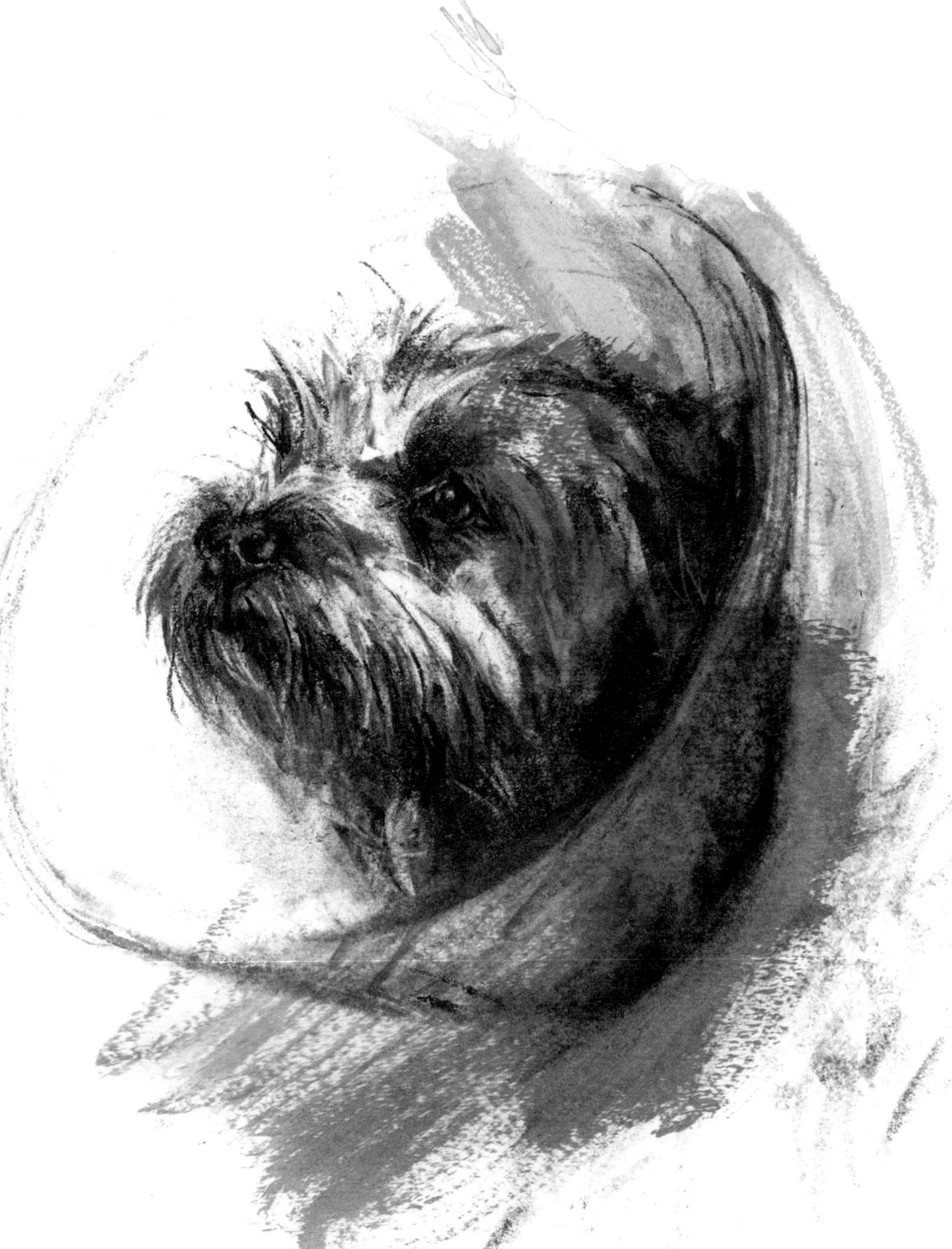

For a while you'd lost your voice.
Then, one day, the sun came out
and your voice came, too.
Your bark lit up my world.
I had mended your broken pieces,
but you did far more than mend mine.
You allowed me to be at peace
with my brokenness.

We should all take time to smell the
daisies and feel the light.

Oh, my beautiful Keira,
you are so right.

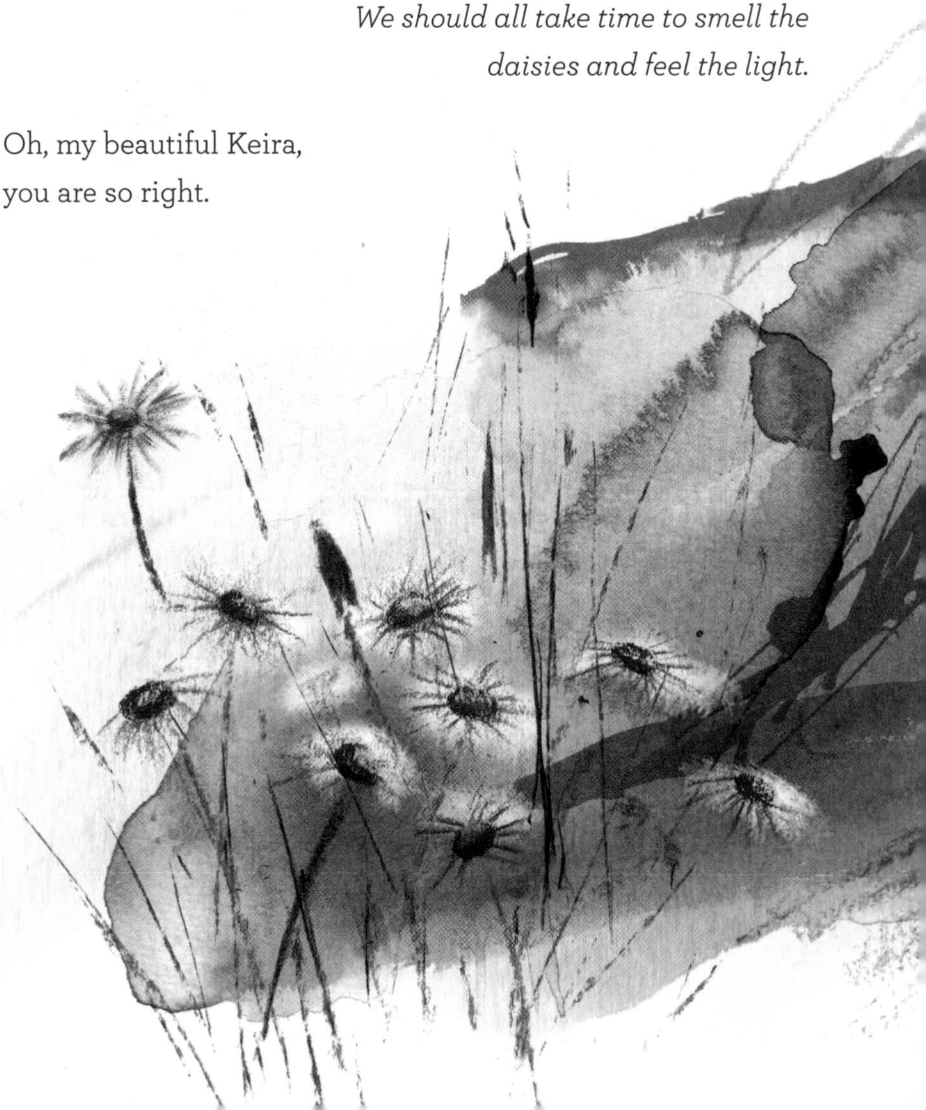

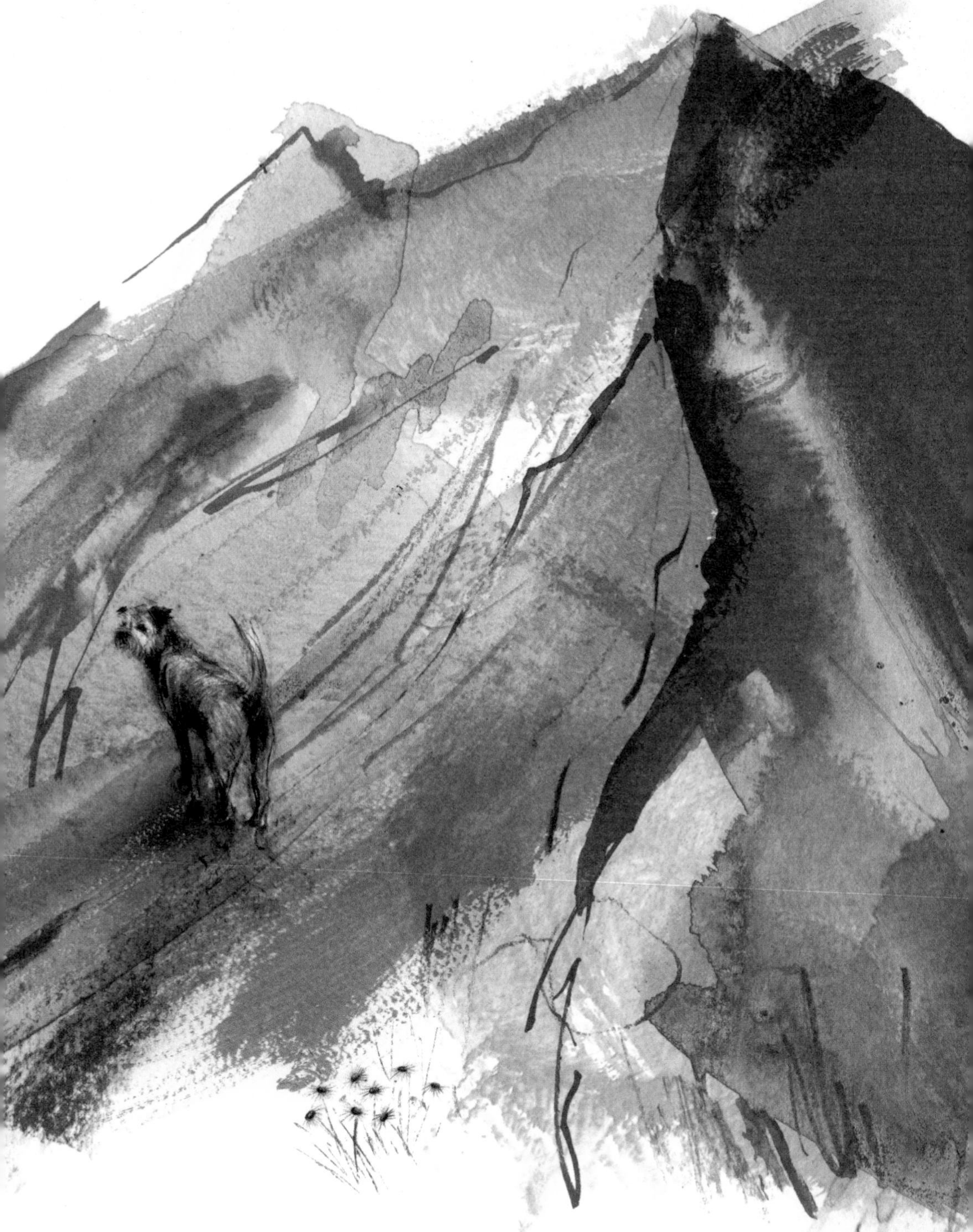

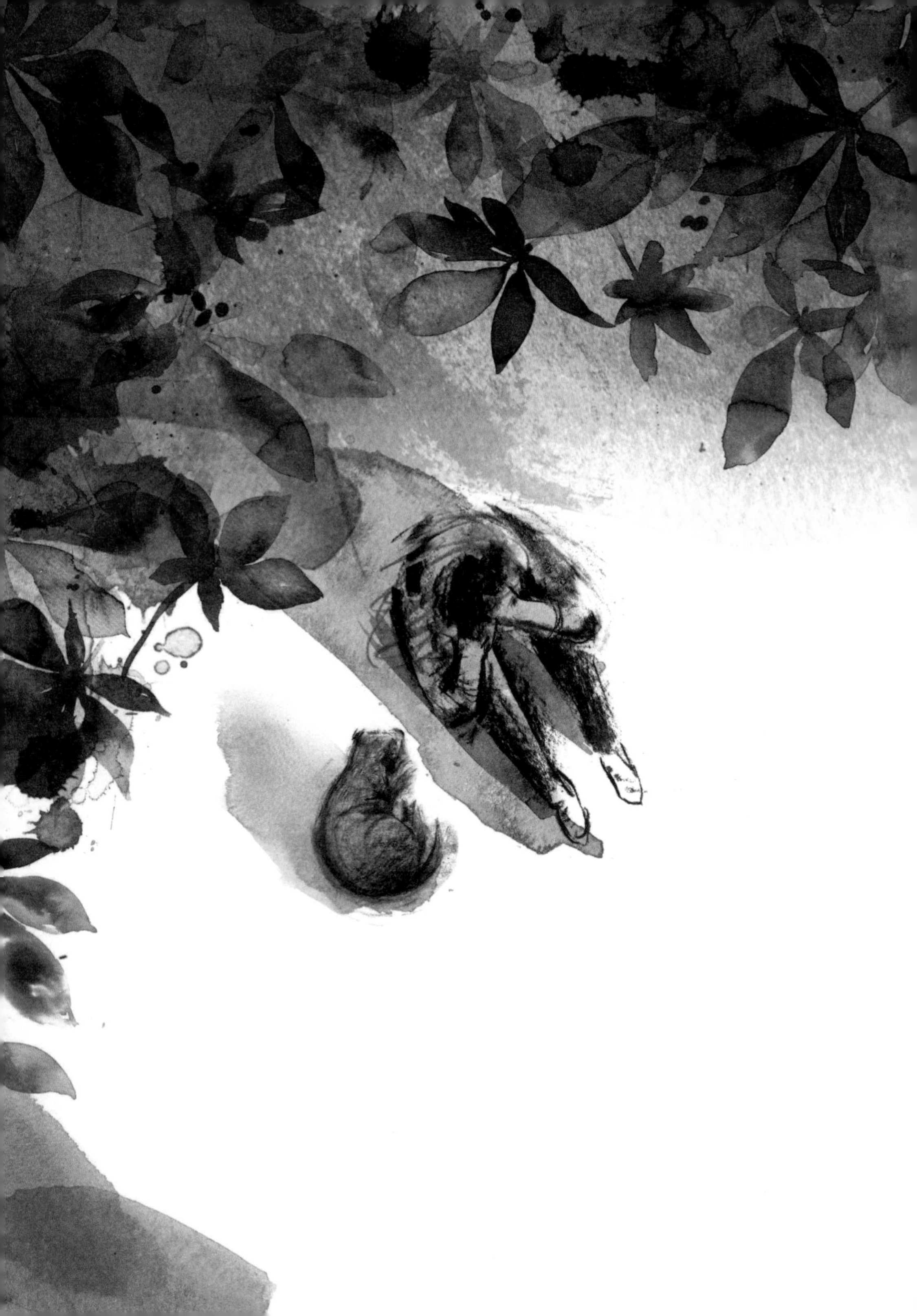

I have always had a problem with regrets.

What are they?

Well – it's when you can't let go because
you think you've made a mistake.

But wherever there is freedom,
there are mistakes.
Mistakes can be blessings.
They can make you
if they don't break you.
You gave me the freedom to
be the best I could be.
Life is just a prison
if you never try at all.
To be afraid of mistakes is
to be afraid of freedom.
To be afraid of freedom is
to be afraid to live.
What else is living for?

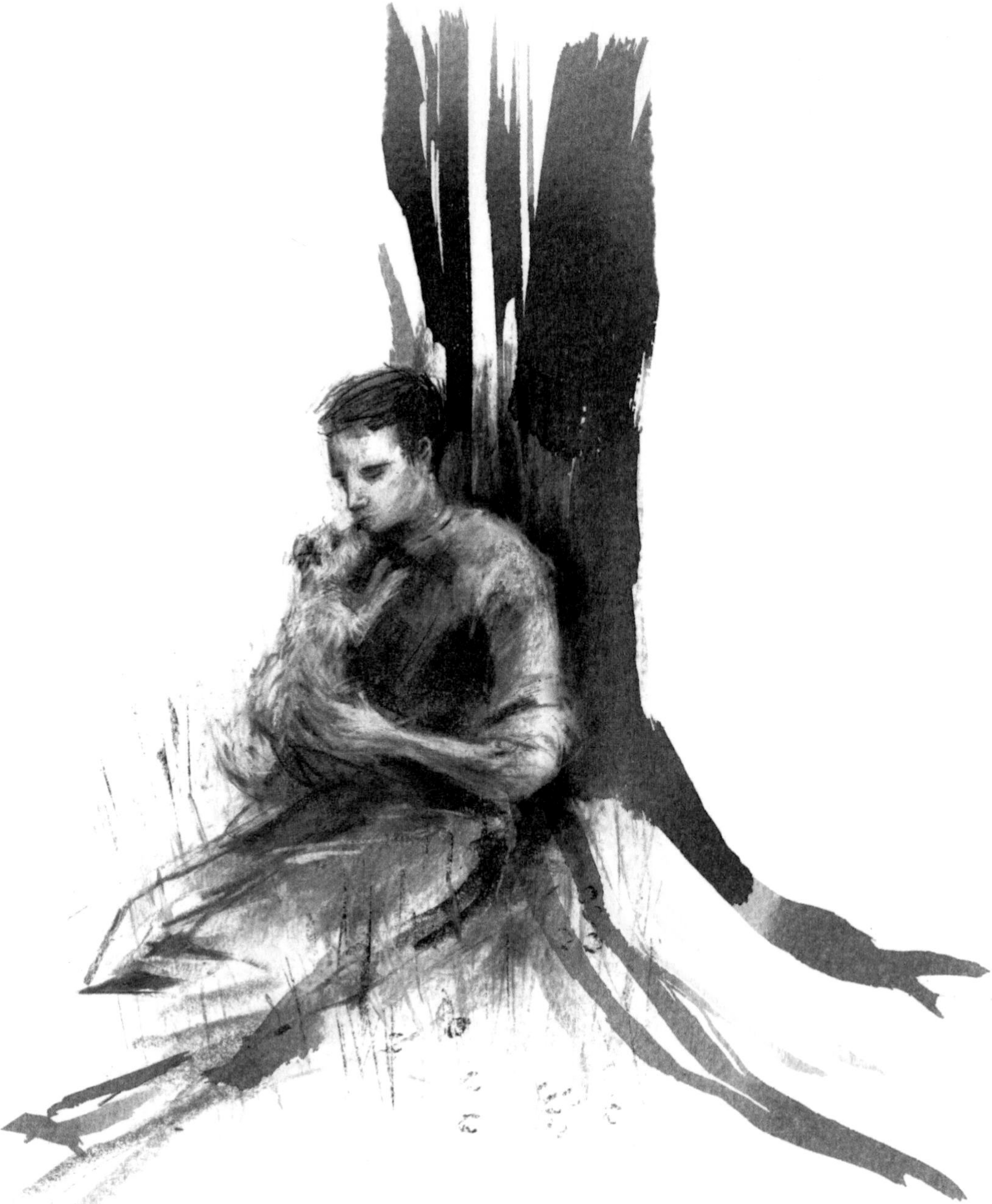

Love is giving someone
freedom to be themselves . . .

And allowing the one you love
to fail sometimes.

All the love in the world
can't stop an accident.

Every success you have ever had
stands on the shoulders of failure.

If we hadn't been through all of that,
we wouldn't be here right now.

And isn't this a lovely place?

Do you feel strong enough for friends?
They've been missing you so much.

Of course! Let them in!

They're bouncier than you, Keira.
You need to be careful.
You might get hurt.

You can never get hurt
when all you want
is the best for each other.

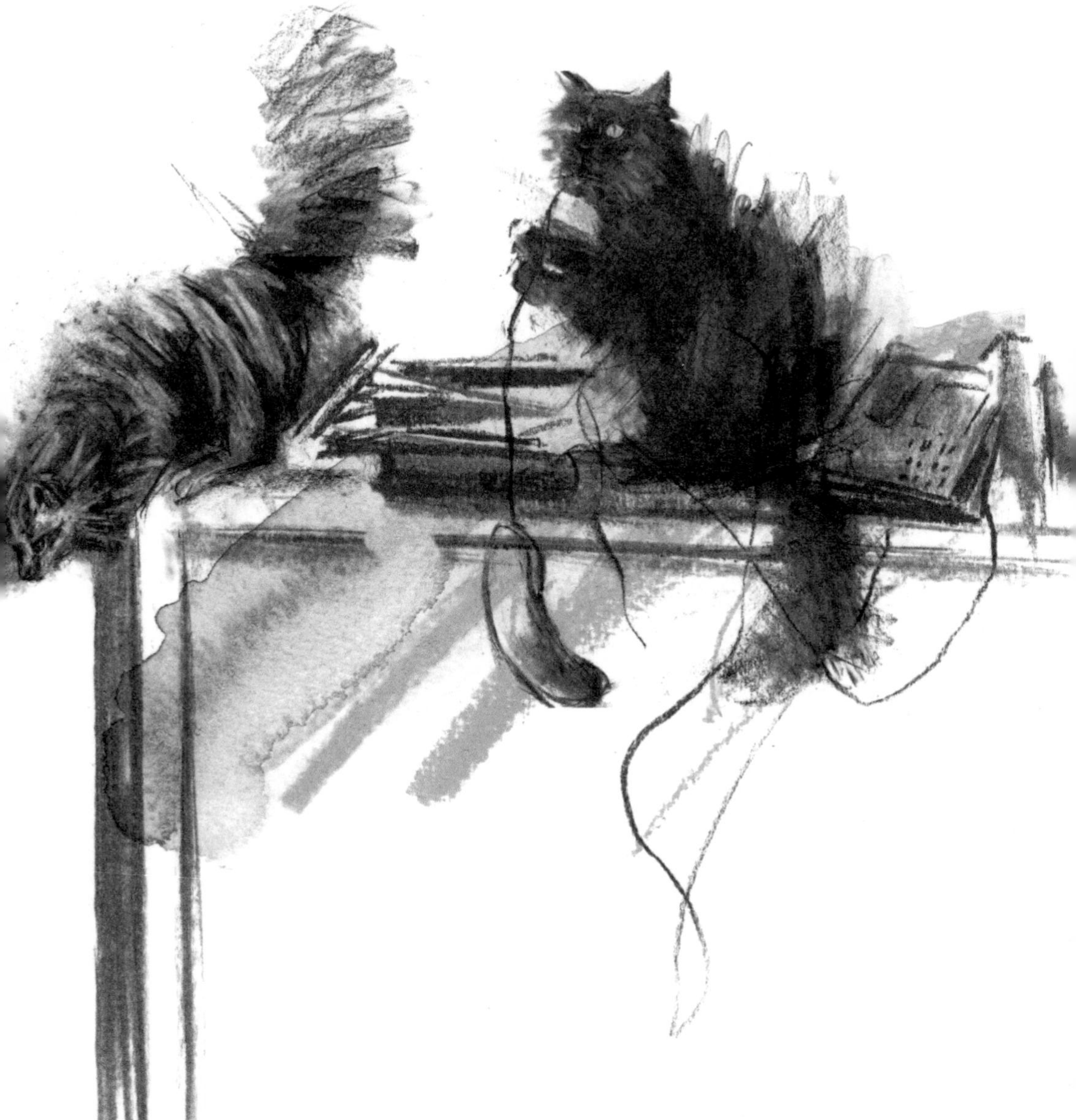

We're getting older, you and I.
We take a slow walk.
It feels rusty
but I can hear you telling me
fear is the rust on dreams.
And you are who you want to become,
not where you came from.

You taught me that to feel worthwhile
you need to spend a *while* on *worth*.
You are worth everything to me
and you showed me that I was enough.

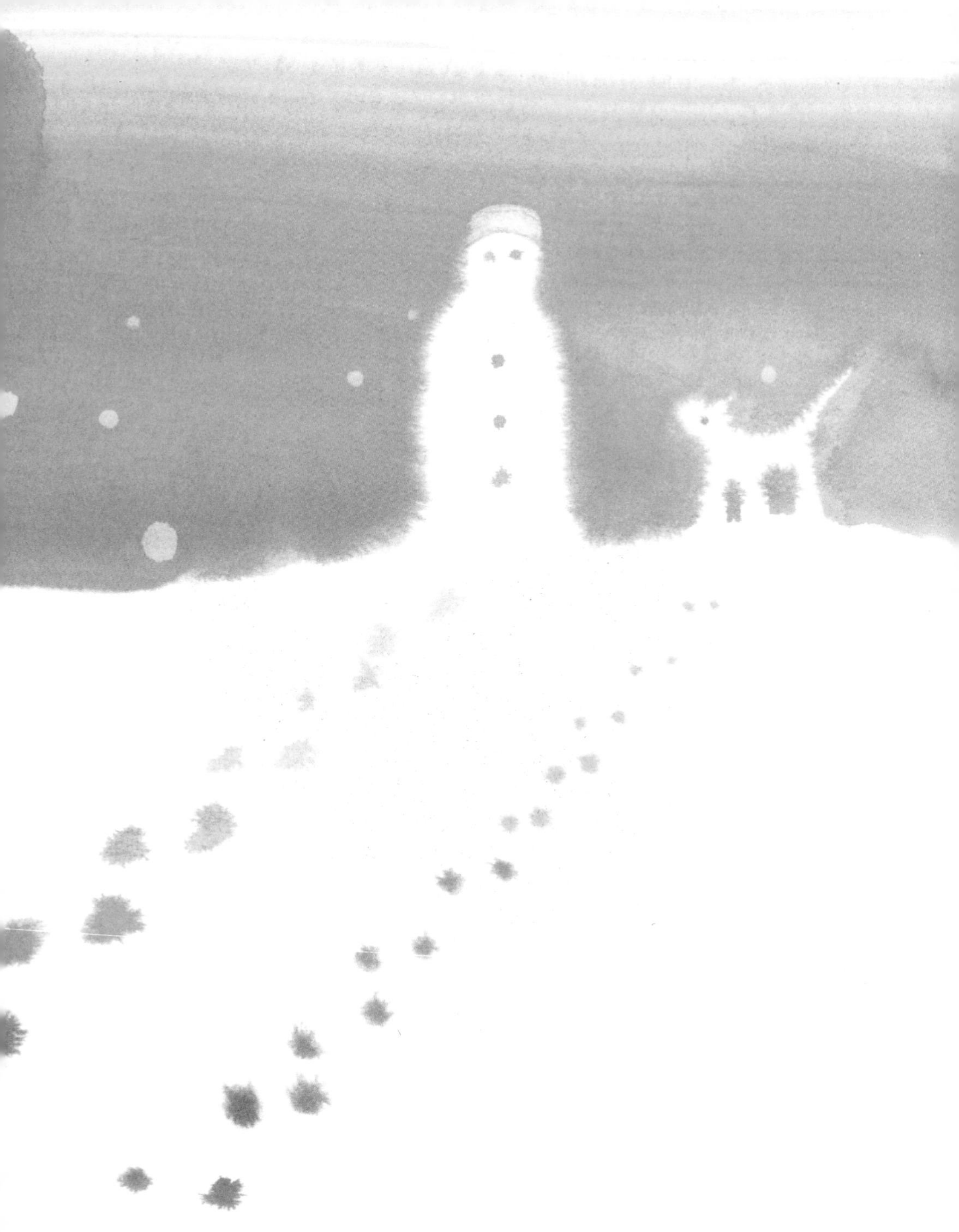

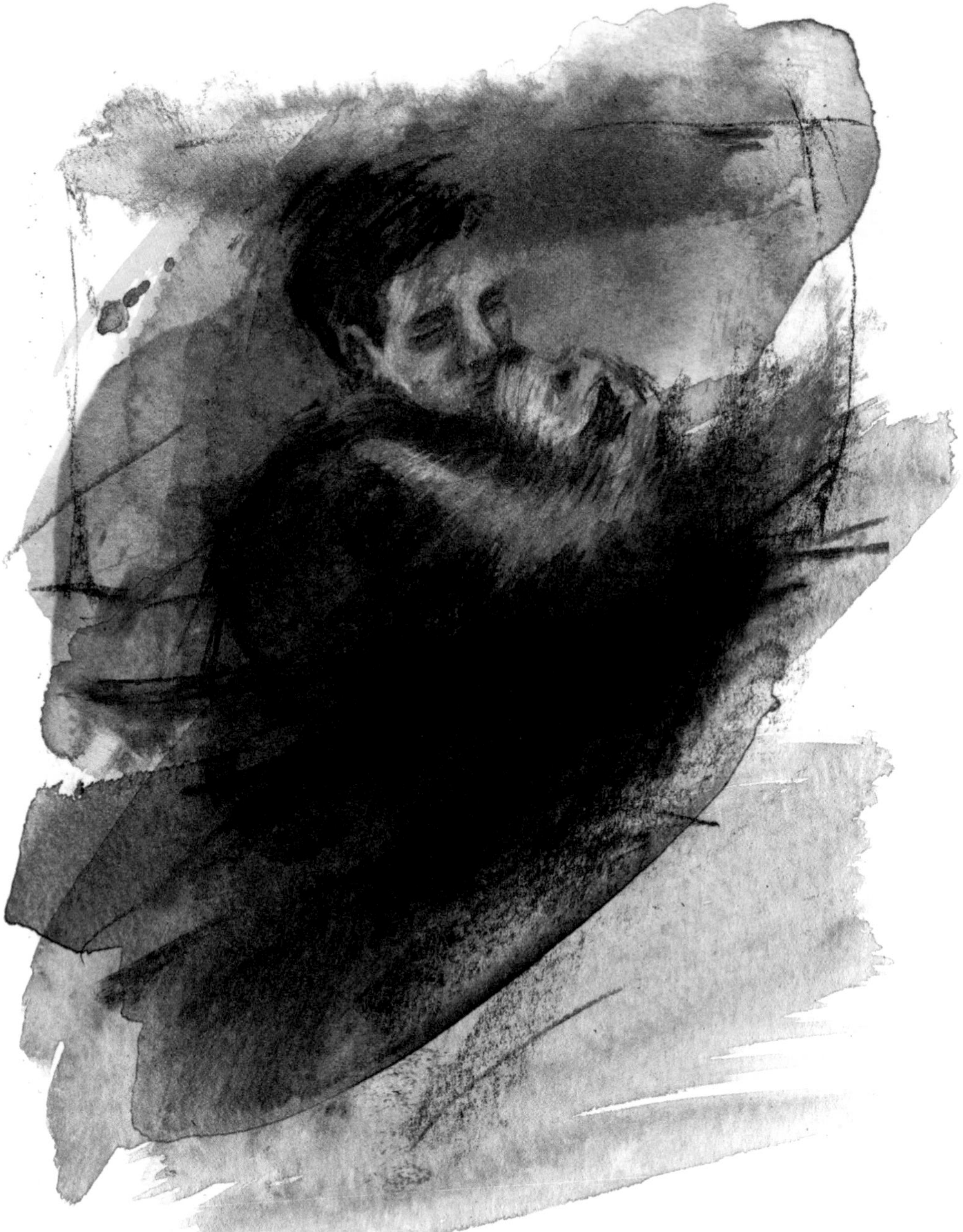

But as your strength returned,
mine was sapped away.
I was so scared.

My mammy was fading fast with her illness
and would soon be gone.
I'd come off the phone to her,
I was crying and you held me like a child.

Bad things happened to me back then.
I didn't know any better,
and so I ran.
Running fast was all I ever knew.
Running on empty for so long.
Life caught up with me when I stopped running.
And you caught me as I tumbled down.

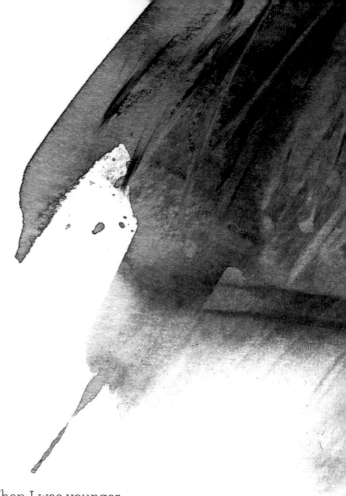

When I was younger,
I wished upon the brightest star in heaven
that I could be more strong, brave and clever,
to help animals, to take away their pain.
You sat by my feet as I studied for exams,
as I borrowed money to build my dreams
and as my hospitals took shape.
Night after night, you sat there,
pawing my knee every now and then,
telling me everything would be okay.
My dreams owe their shape to you.

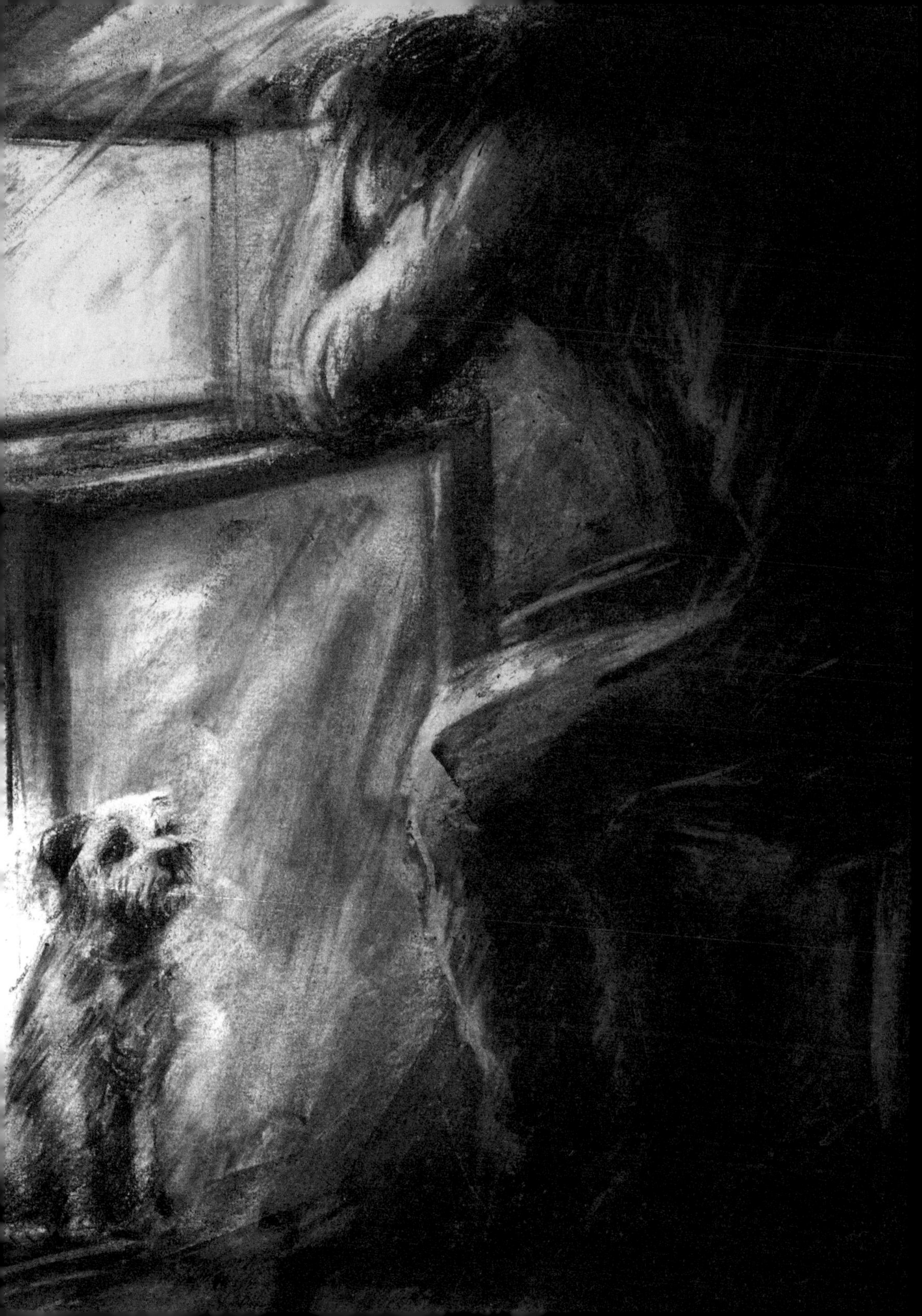

I thought that if I gave all of my heart,
all of the time, that would be enough.
But this wasn't the world we signed up for any more.
Morals muddled by money.
Love valued only when it is lucrative.

I lost respect for those I held in high esteem.
I lost trust in those I thought I could rely on.
They wanted what I gave them,
but they didn't want me.
Everybody has a price, they say.
But I wasn't for sale.
You understood and stood by my side.

I couldn't be a cog in their machine.
Most mean well and deserve to choose their path.
But I was broken and my dream had been stolen.

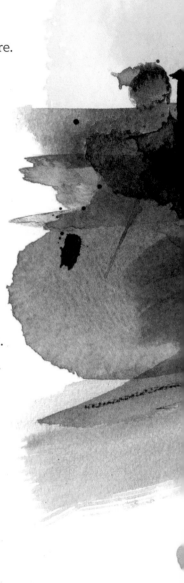

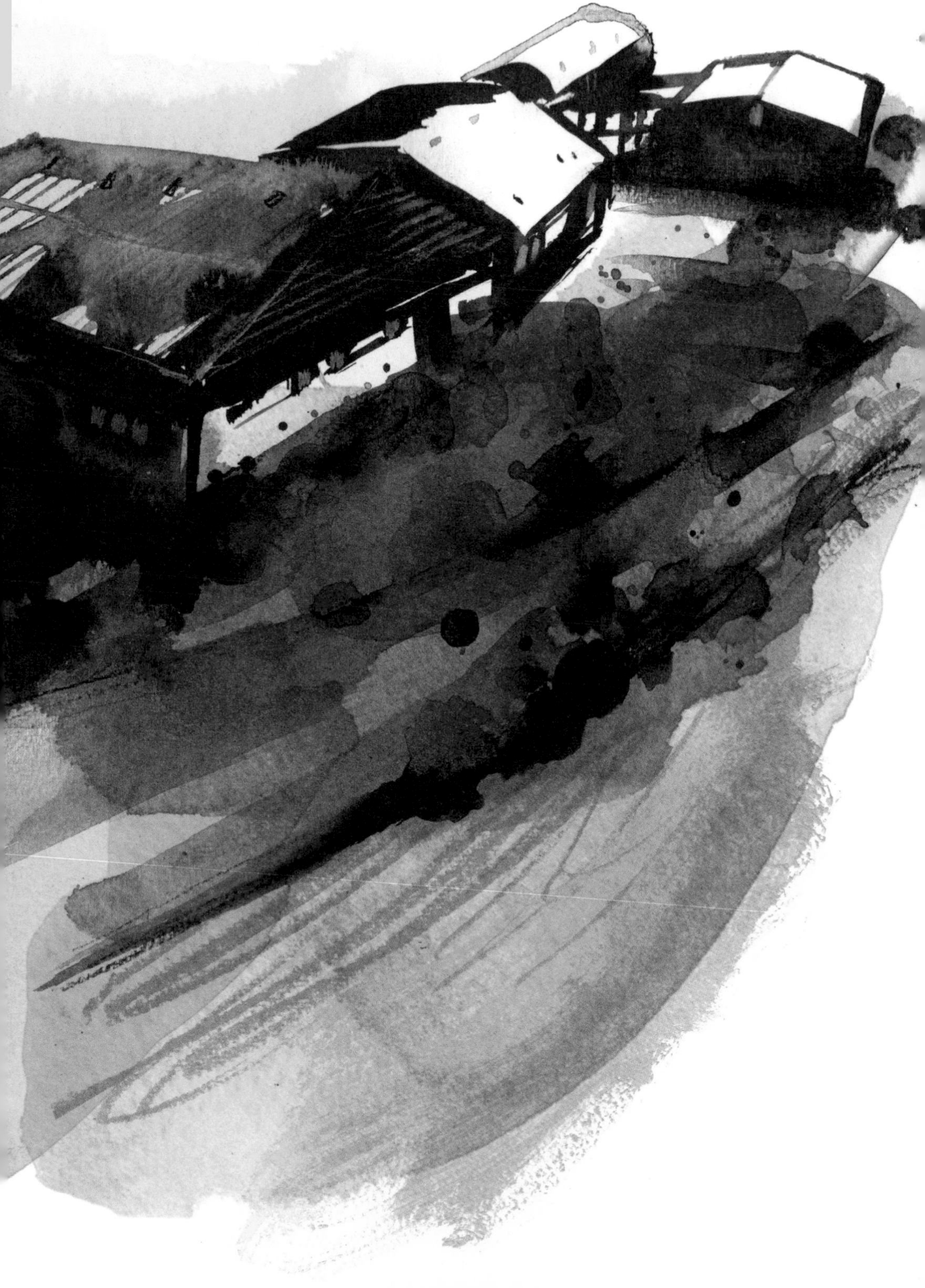

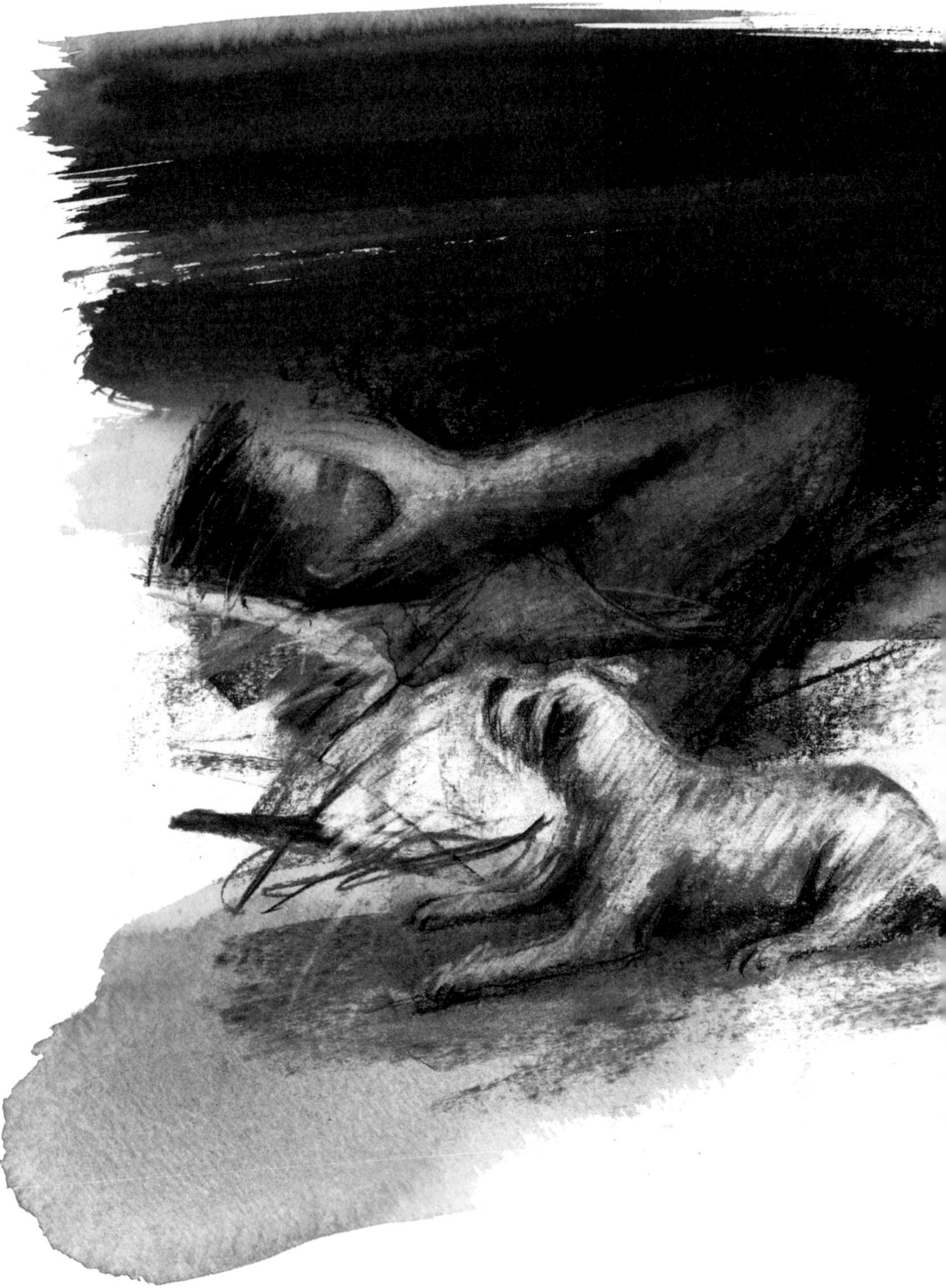

Sleep, when it came, was never enough.
The day's worries bled into nightmares.
Darkness eating me from inside out.
A candle burning from both ends.

I questioned why I existed at all.
Just banging my head on a wall.

I'm in pain, Keira.
This pain is different. I can't reach it.

I can feel your heart, like it's my own.
Let me try to take your pain away,
as you did mine.
Maybe we can dream bigger
than ever before.

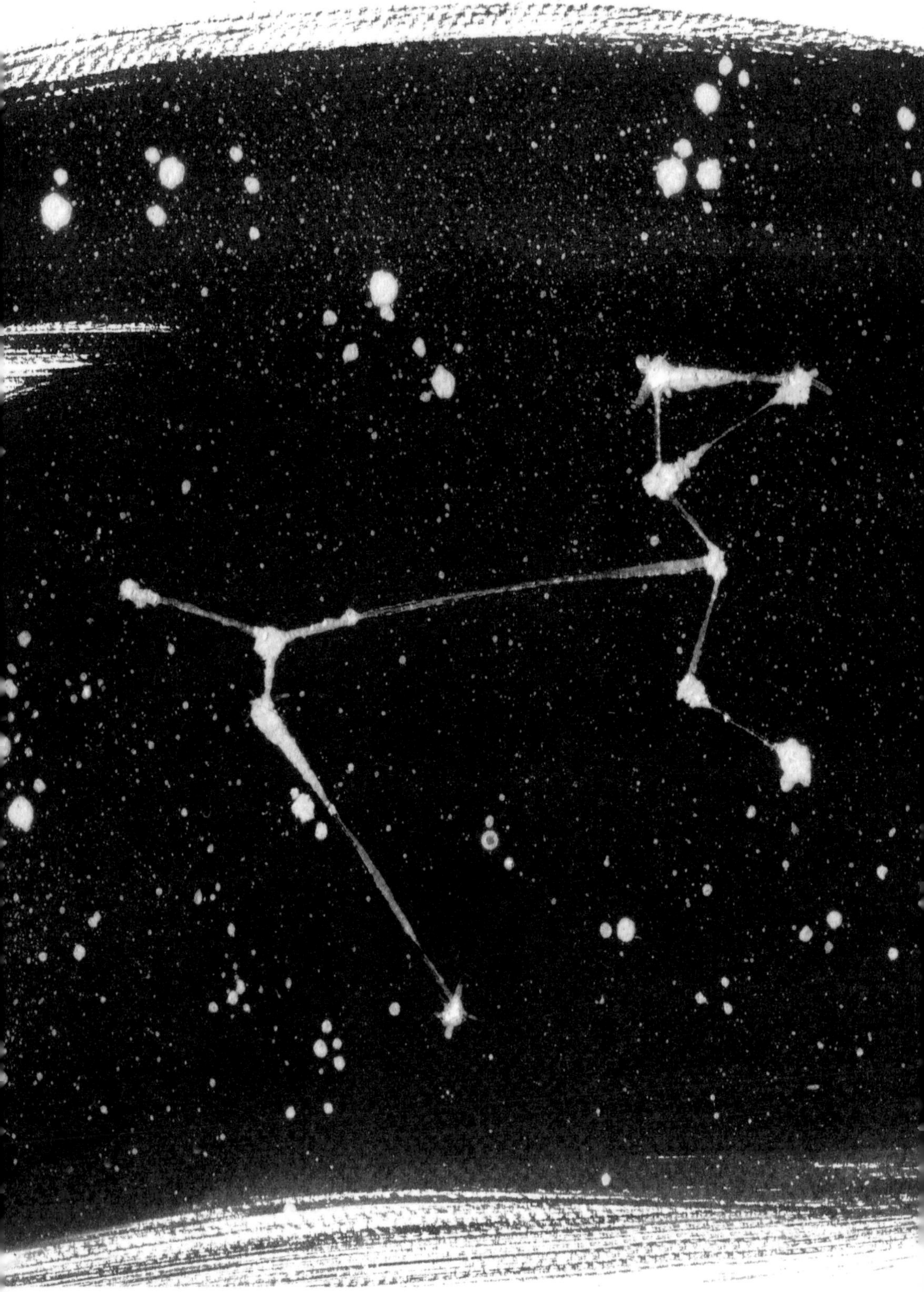

I was fine on the outside,
but inside I was screaming.
And only you could hear.
You showed me that I would never find
my truth through my purpose,
only my purpose through my truth.
You were my great awakening.
My heart was locked.
I needed to open the door,
but I couldn't do it alone.
You held the key in your mouth,
each and every time,
when my heart was in my own.

Your wagging tail was my compass,
and your cuddles my comfort.
Your paws held me up,
as hands all around dragged me down.

I felt so small.
But you showed me
that only the small things can
change anything at all.

If I sought meaning outside,
I would feel meaningless inside.

A hole had opened up inside me, and I fell in.
You filled that hole with love,
and made me whole again.

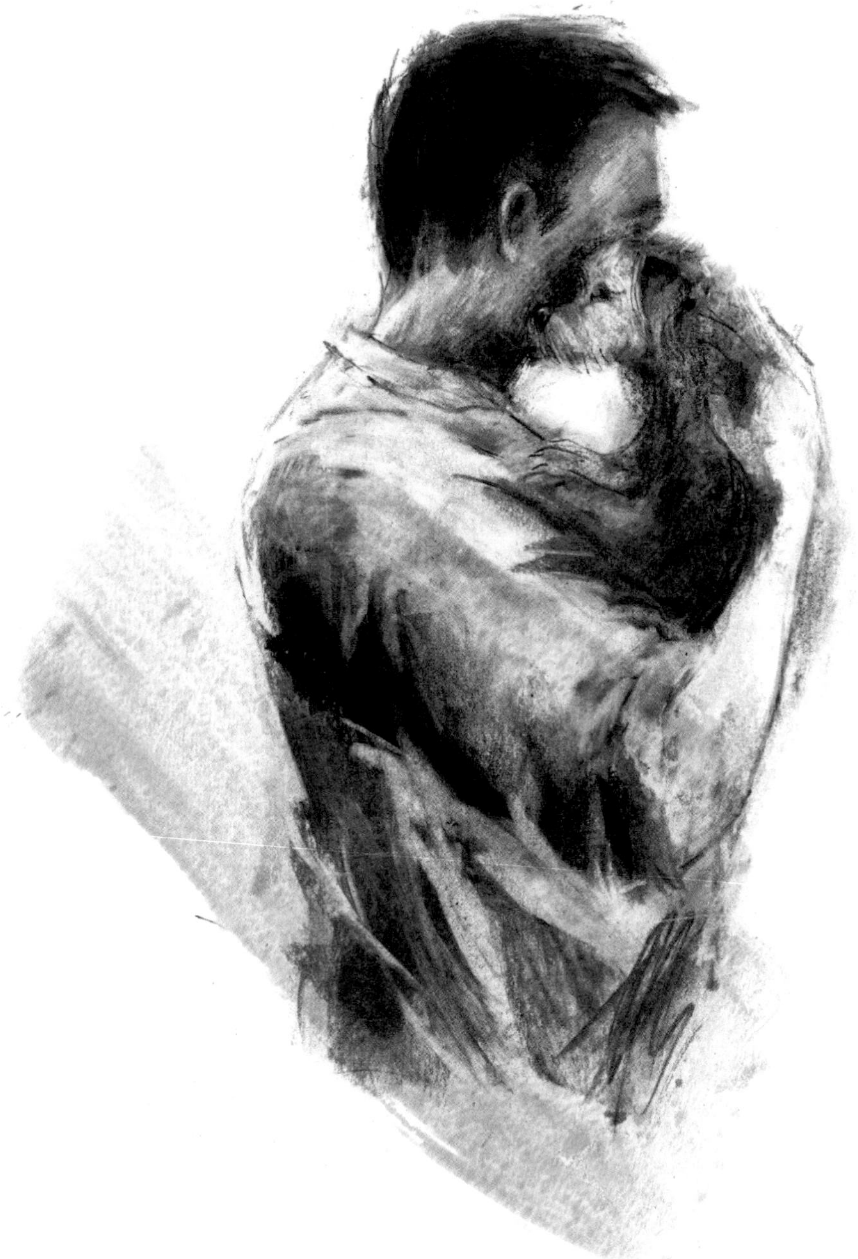

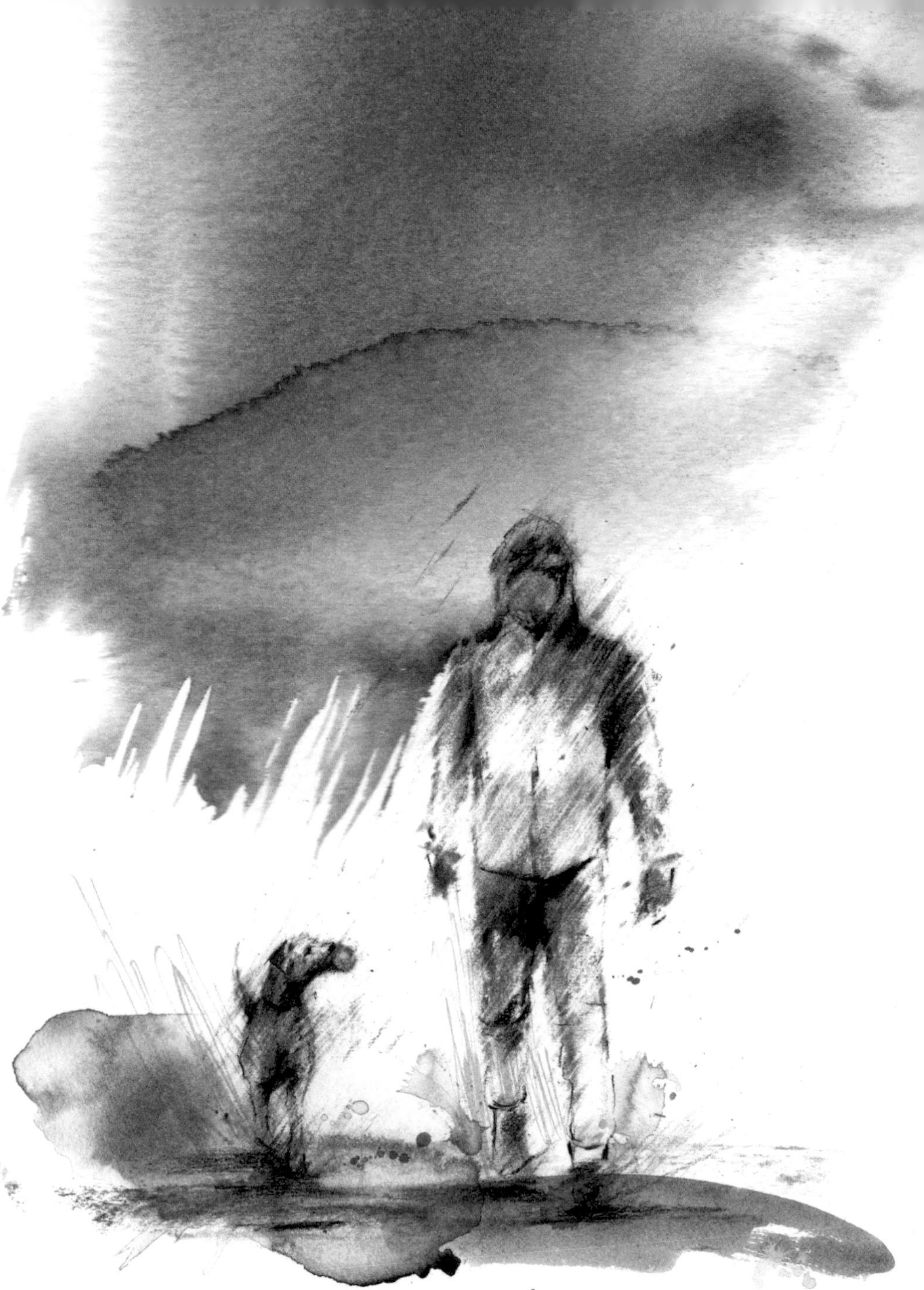

Unconditional love
doesn't tell you what to think,
but rather reaches inside you
and asks you to add something
to make the flame brighter,
something only your light can add.

You taught me how you can only *find* yourself,
if you can find a way to *love* yourself
and forgive yourself for what has passed.
You bandaged my pain with your love.
And helped me see the value in everything,
even a soggy tennis ball in the rain.

Admitting I was weak
became my superpower.

With you by my side,
knowing that I didn't have all the answers
and that I didn't need to,
just doing my best was the best thing to do.

Your patient joy taught me that
the only way to banish the darkness of my past
is to be light in the present.

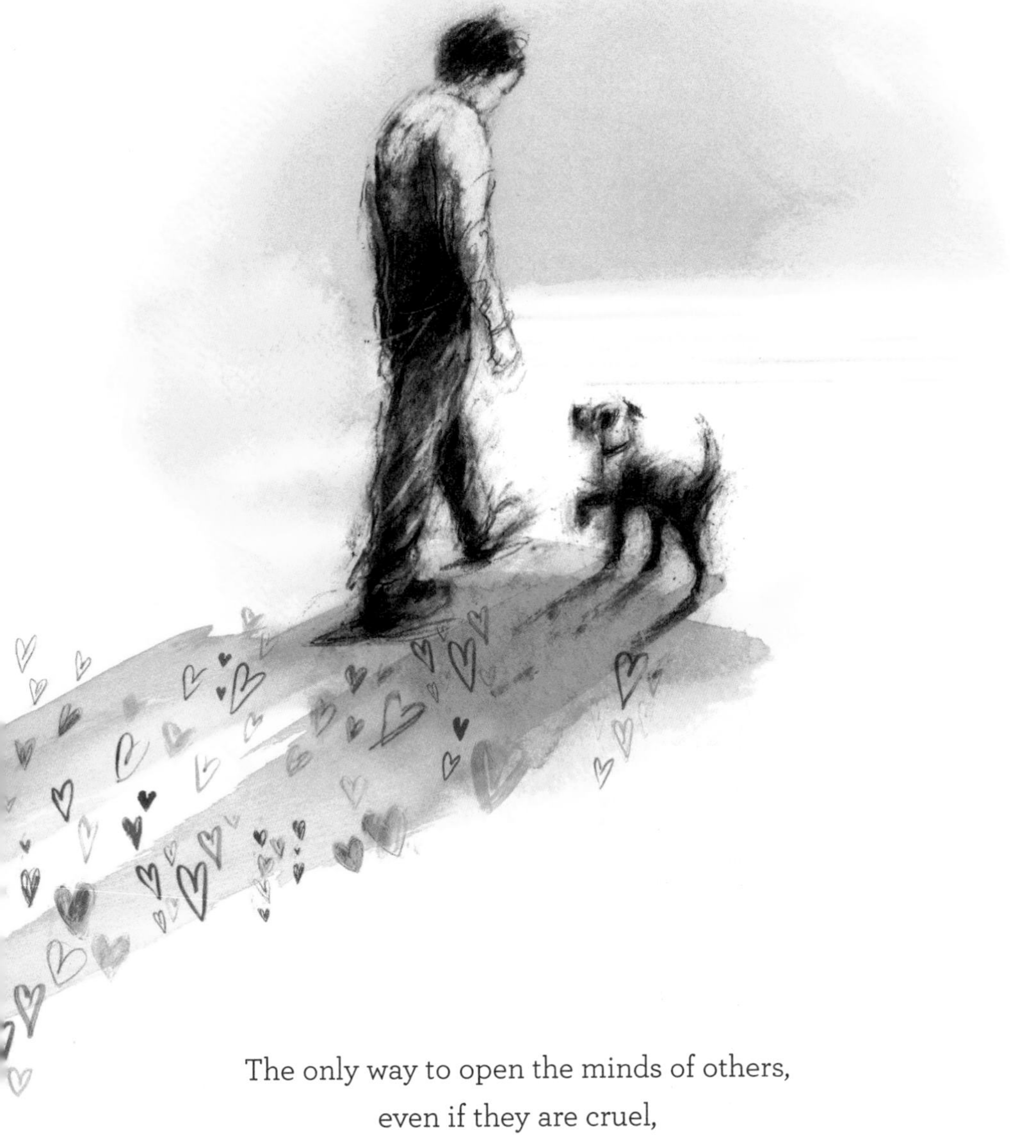

The only way to open the minds of others,
even if they are cruel,
is to open your own heart
and send them love and kindness regardless.

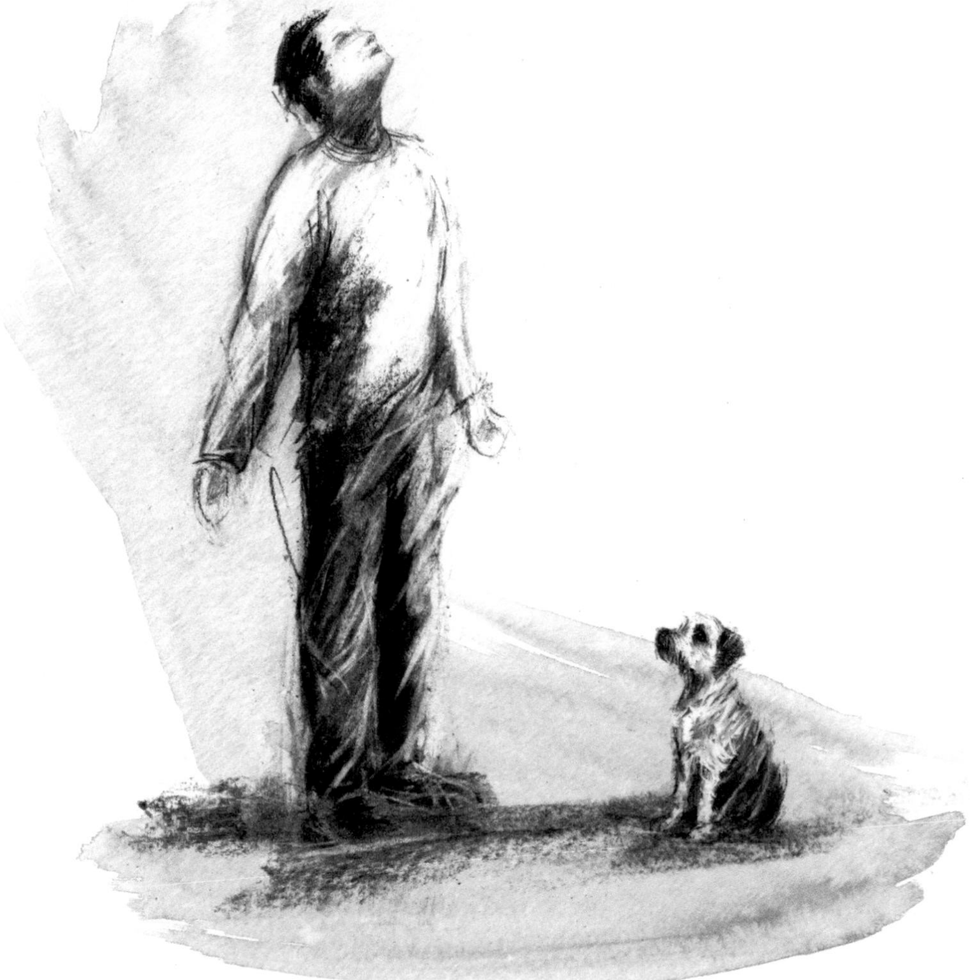

You finally stopped me running
when you couldn't run yourself.
Your love kept me grounded
and lifted me up.
You were my parachute when I jumped
and my cushion when I fell.

Broken, but ready to heal.
Hurt, but ready to forgive.
I surrendered to my fears,
trying to trust again.

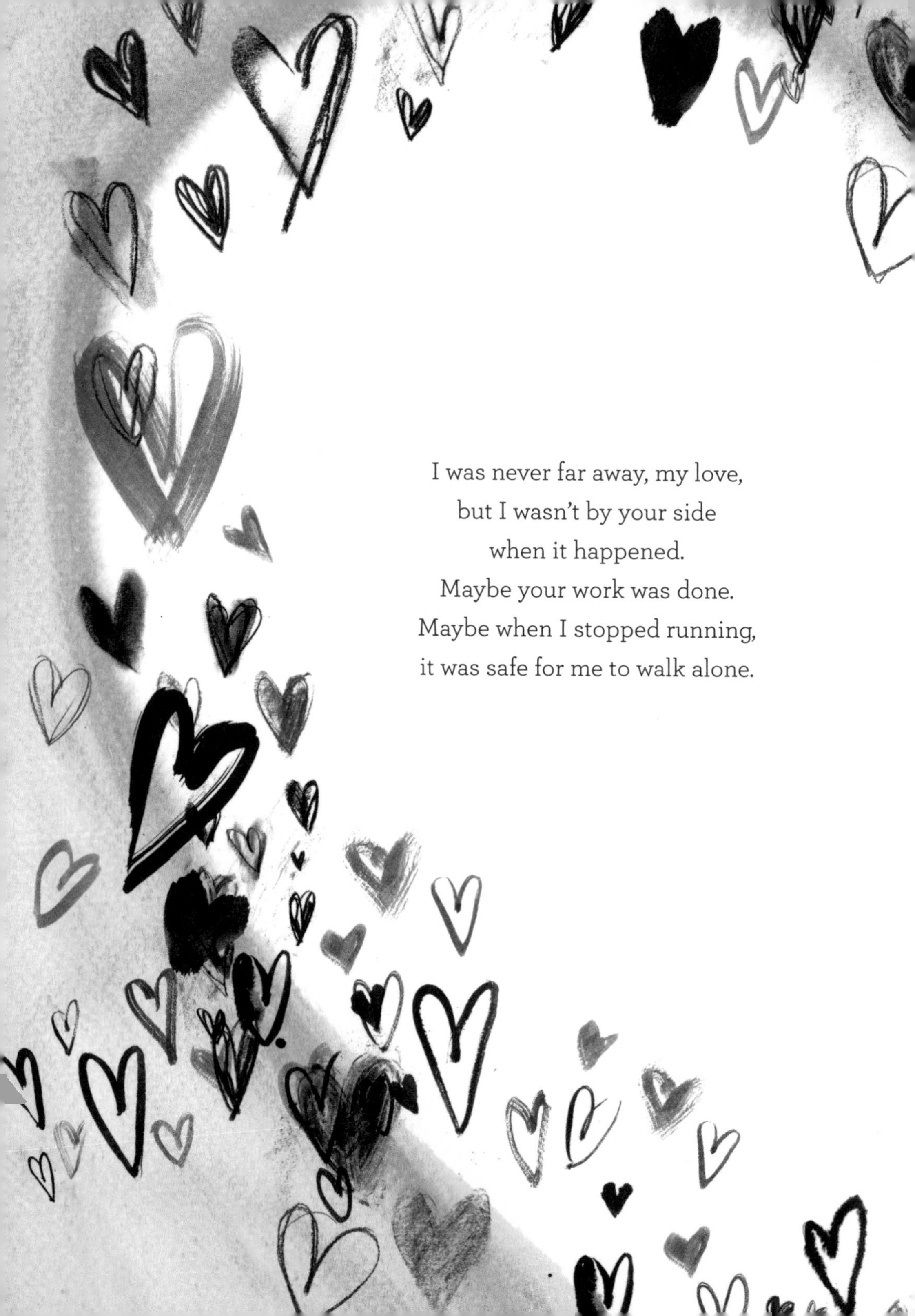

I was never far away, my love,
but I wasn't by your side
when it happened.
Maybe your work was done.
Maybe when I stopped running,
it was safe for me to walk alone.

You had given so much of your heart
the battery had run down.

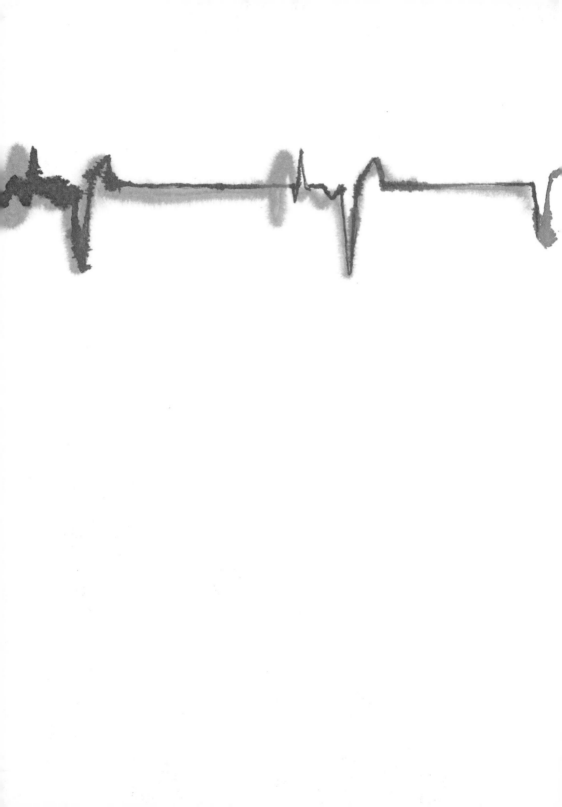

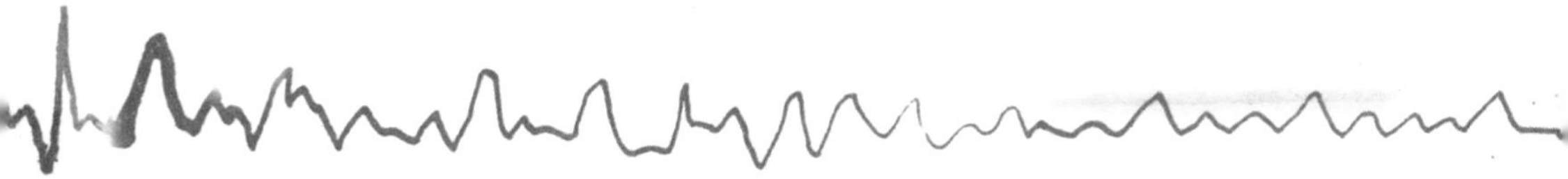

They rushed you to me,
the light already ebbing from your eyes.
Stethoscope.
Lub, dub . . . lub, dub . . .
slow, drifting . . .
dub . . .
Your lungs empty.
I pressed on your chest.
Nothing.

Nothing.

I picked up a blade and I cut down.
I fought for you,
as you fought for me.

I reached into your chest
and felt your heart
like it was my own.

It stopped beating in my hand.
I wanted mine to stop, too.

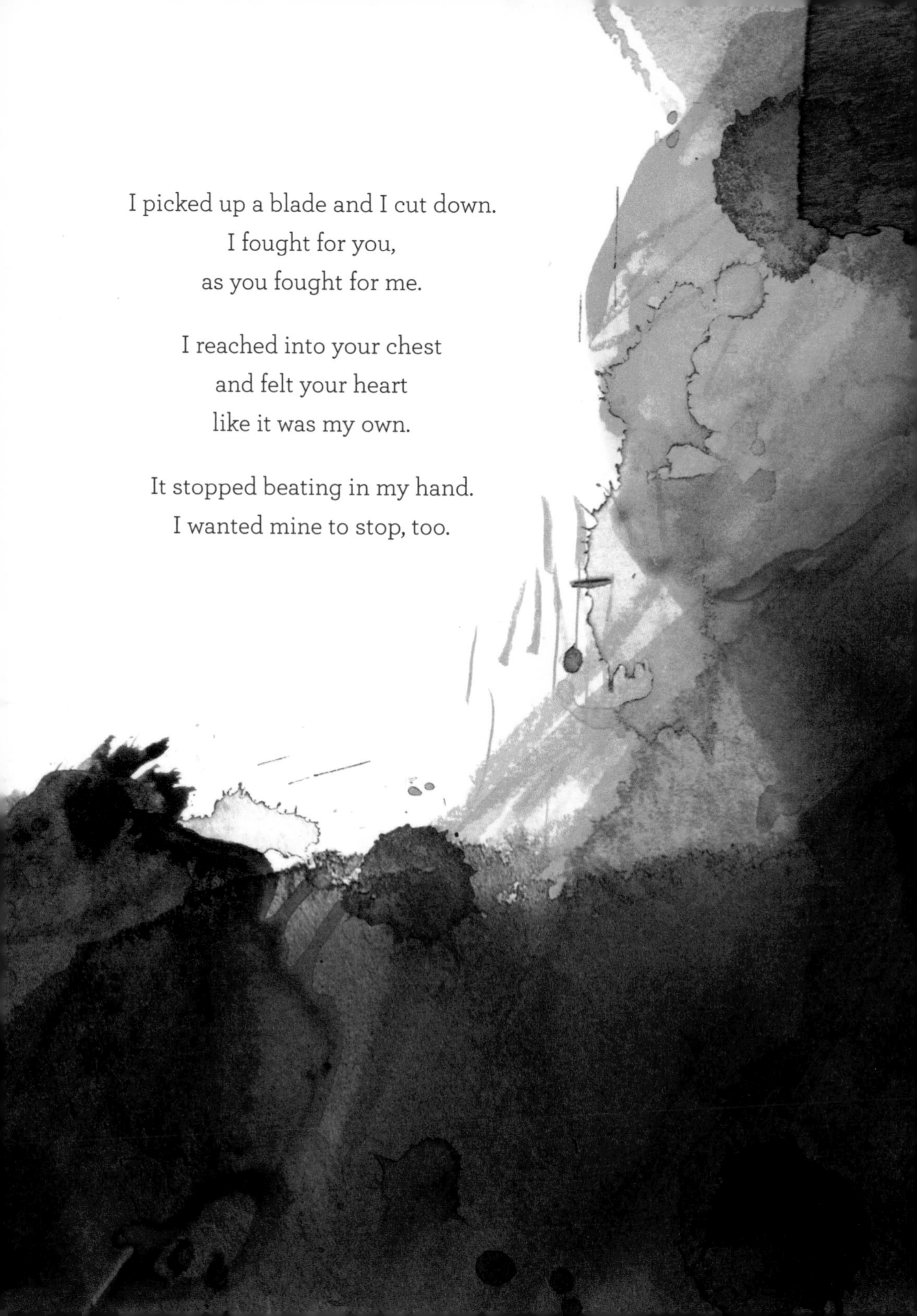

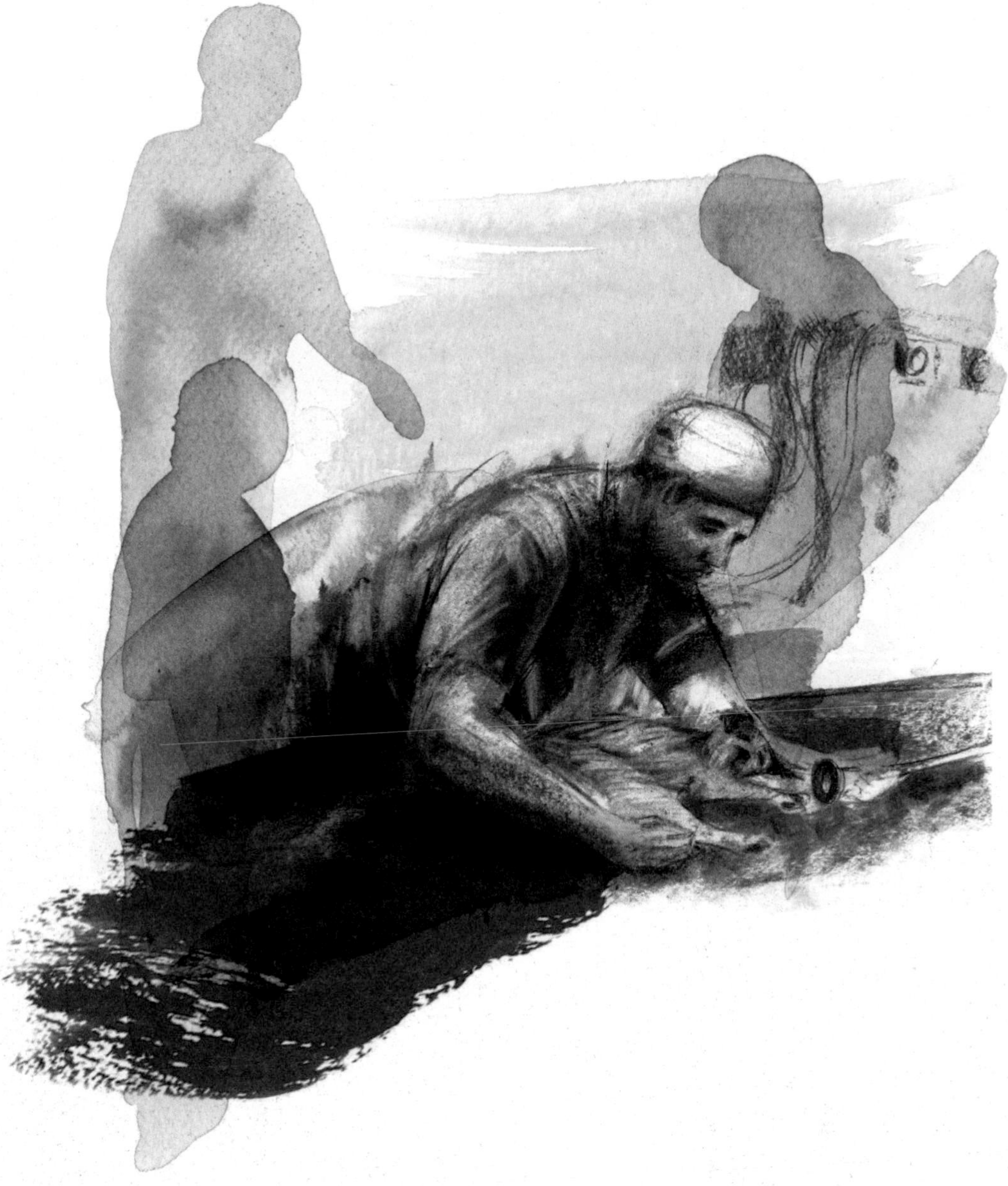

One by one, the bodies drifted away.
A soft hand on my shoulder.
A gentle word in my ear.
I felt and heard nothing.
Slumped over your lifeless body,
crying into a gaping hole
that could never be filled.

I scooped your fragile stardust in my arms.
Hunched and crumpled,
bleeding tears,
I carried you on our final walk.

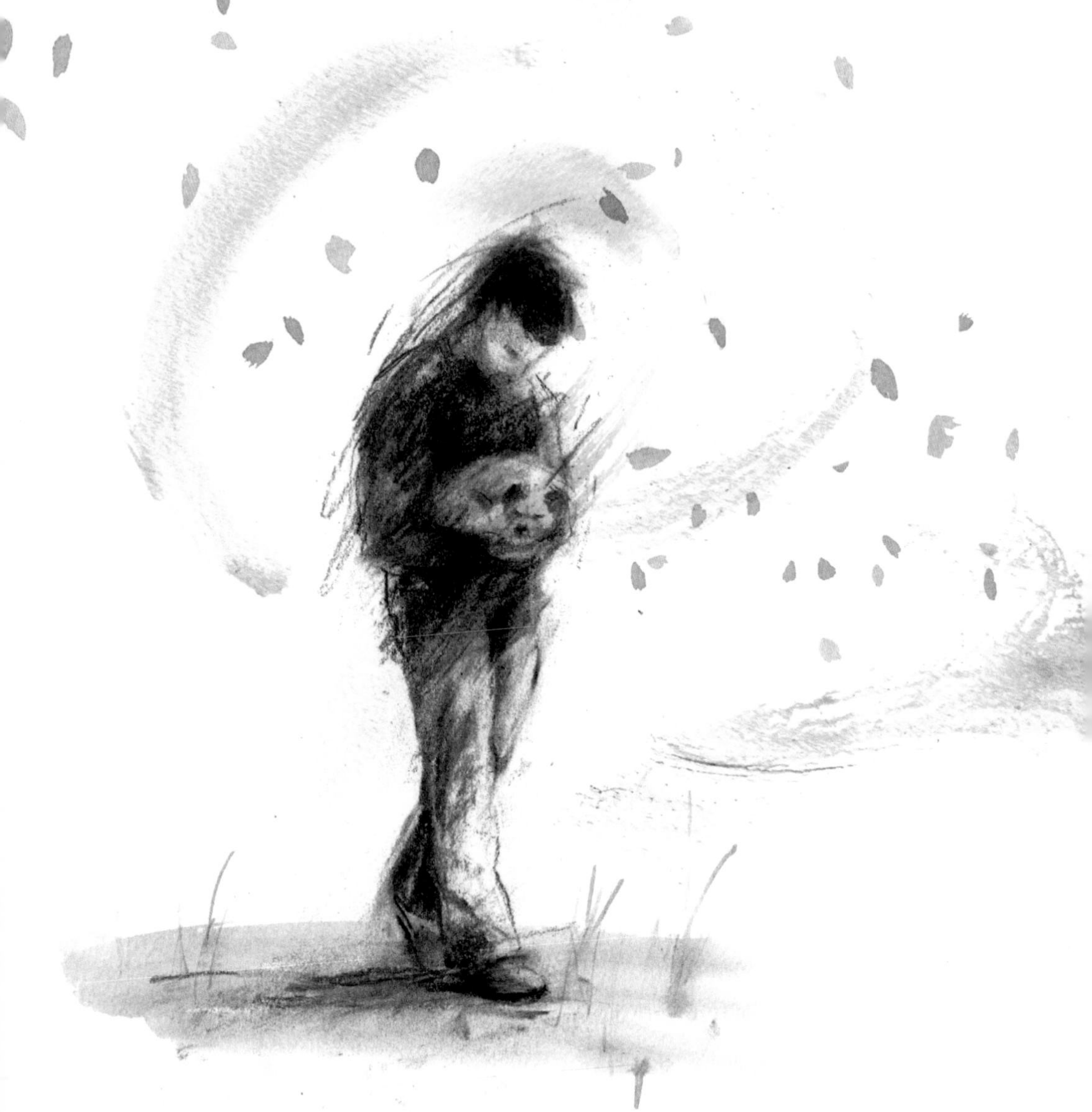

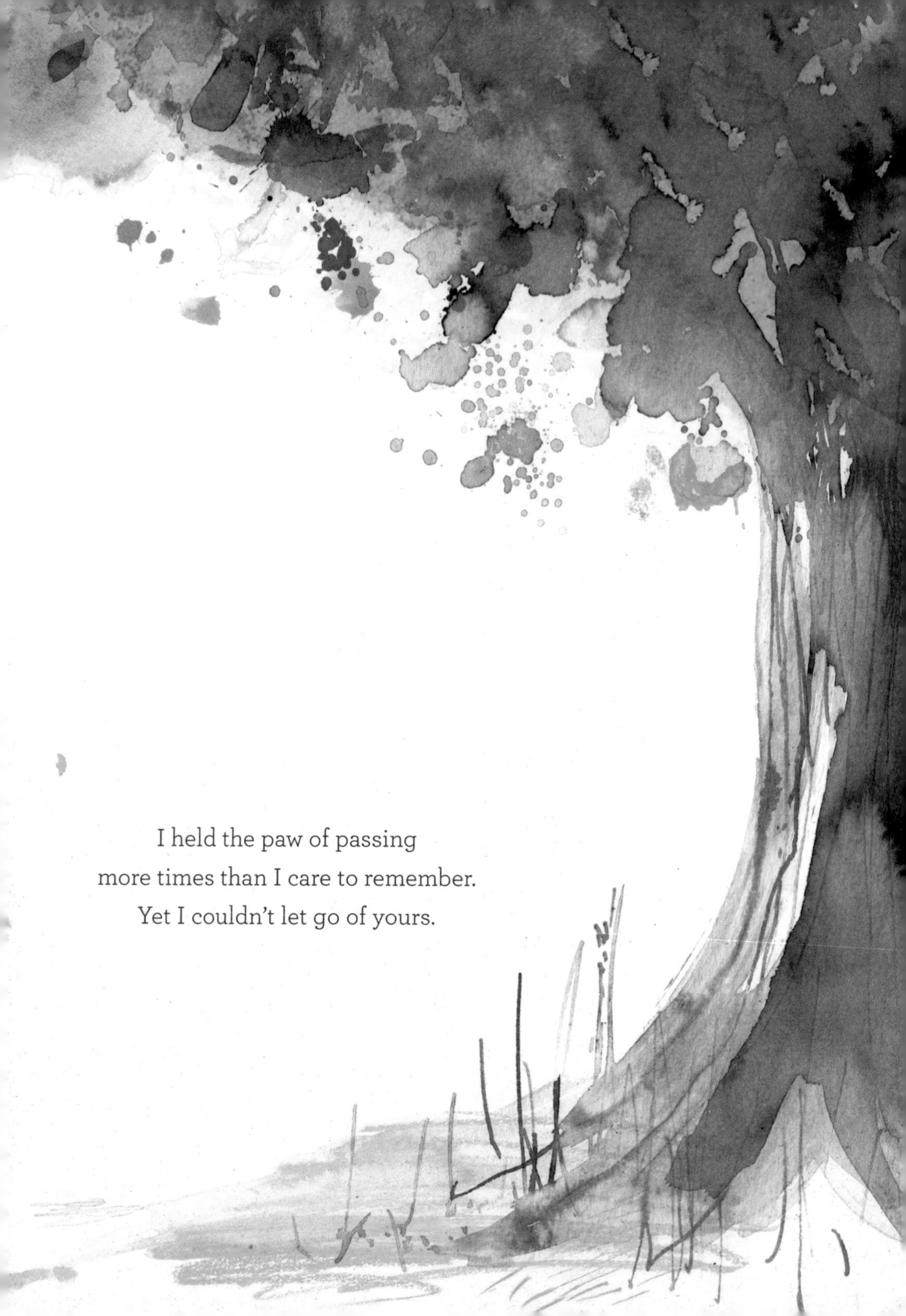

I held the paw of passing
more times than I care to remember.
Yet I couldn't let go of yours.

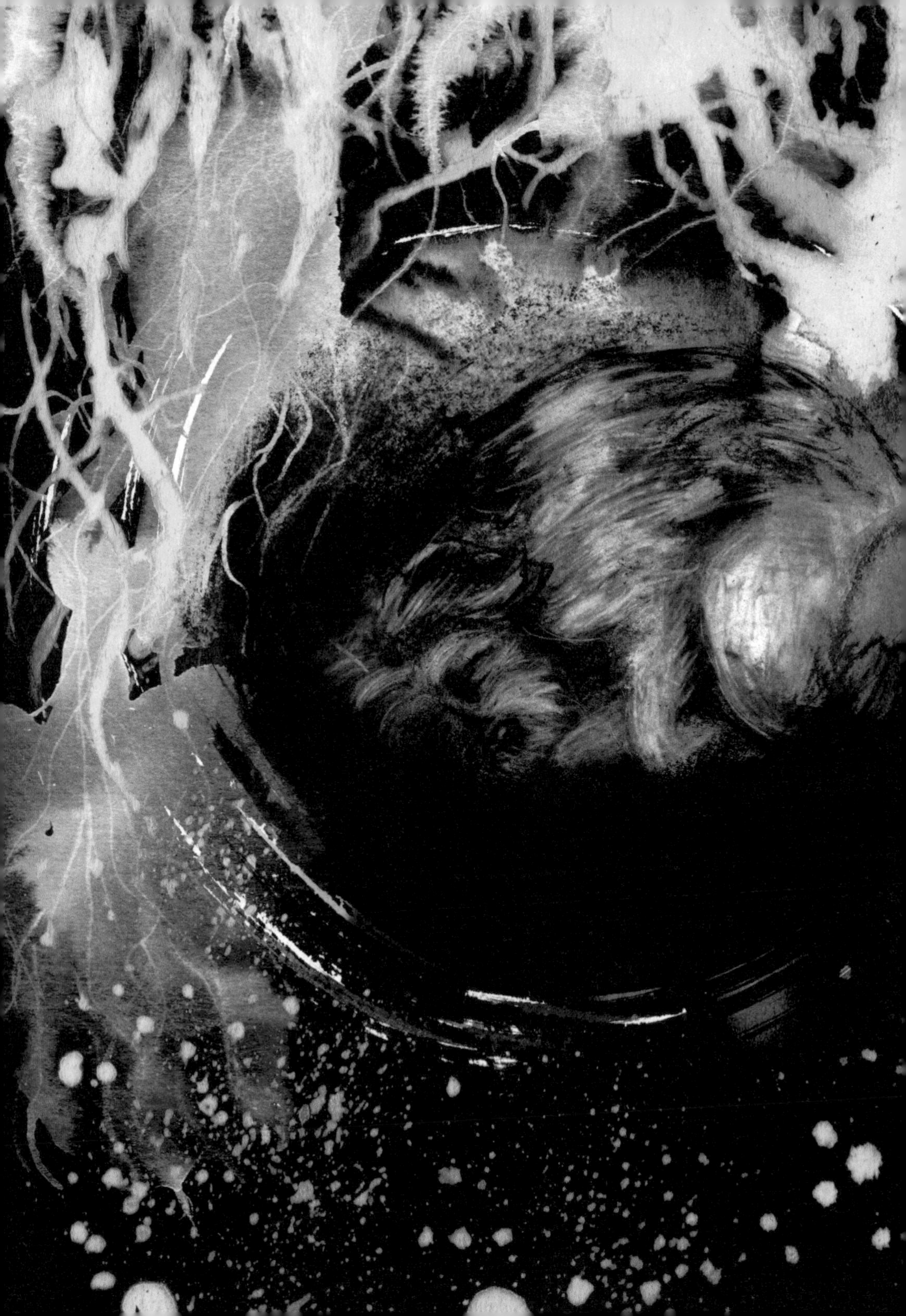

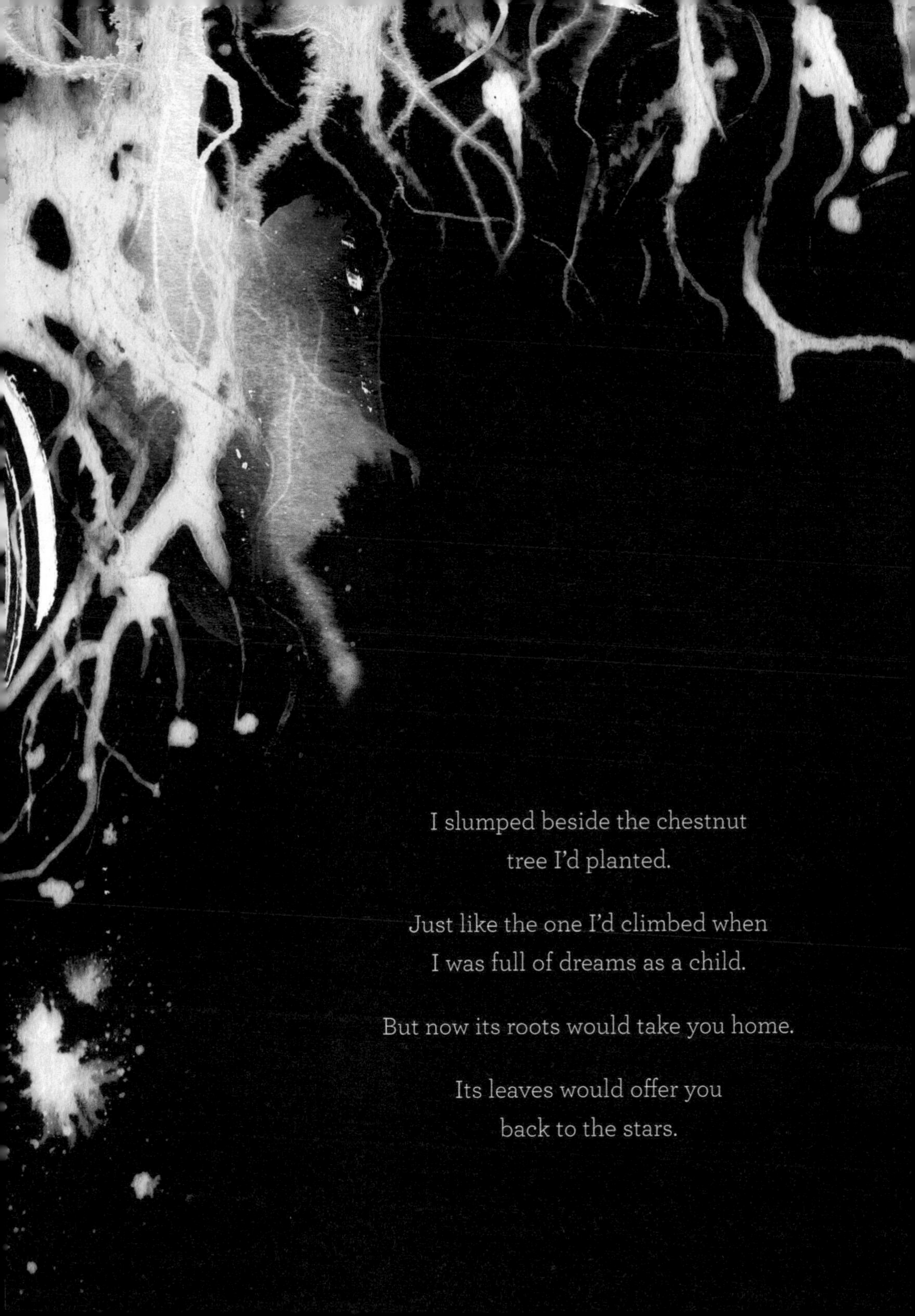

I slumped beside the chestnut
tree I'd planted.

Just like the one I'd climbed when
I was full of dreams as a child.

But now its roots would take you home.

Its leaves would offer you
back to the stars.

I held you for the longest time.
I would need to dig the cold clay,
but I longed to keep you warm.

I could hear your whisper, even then,
bringing solace as I cradled you.
I must now dream as if our hearts still beat as one.

You left me your eternal hope
that tomorrow may be a better day
because you're still in it.

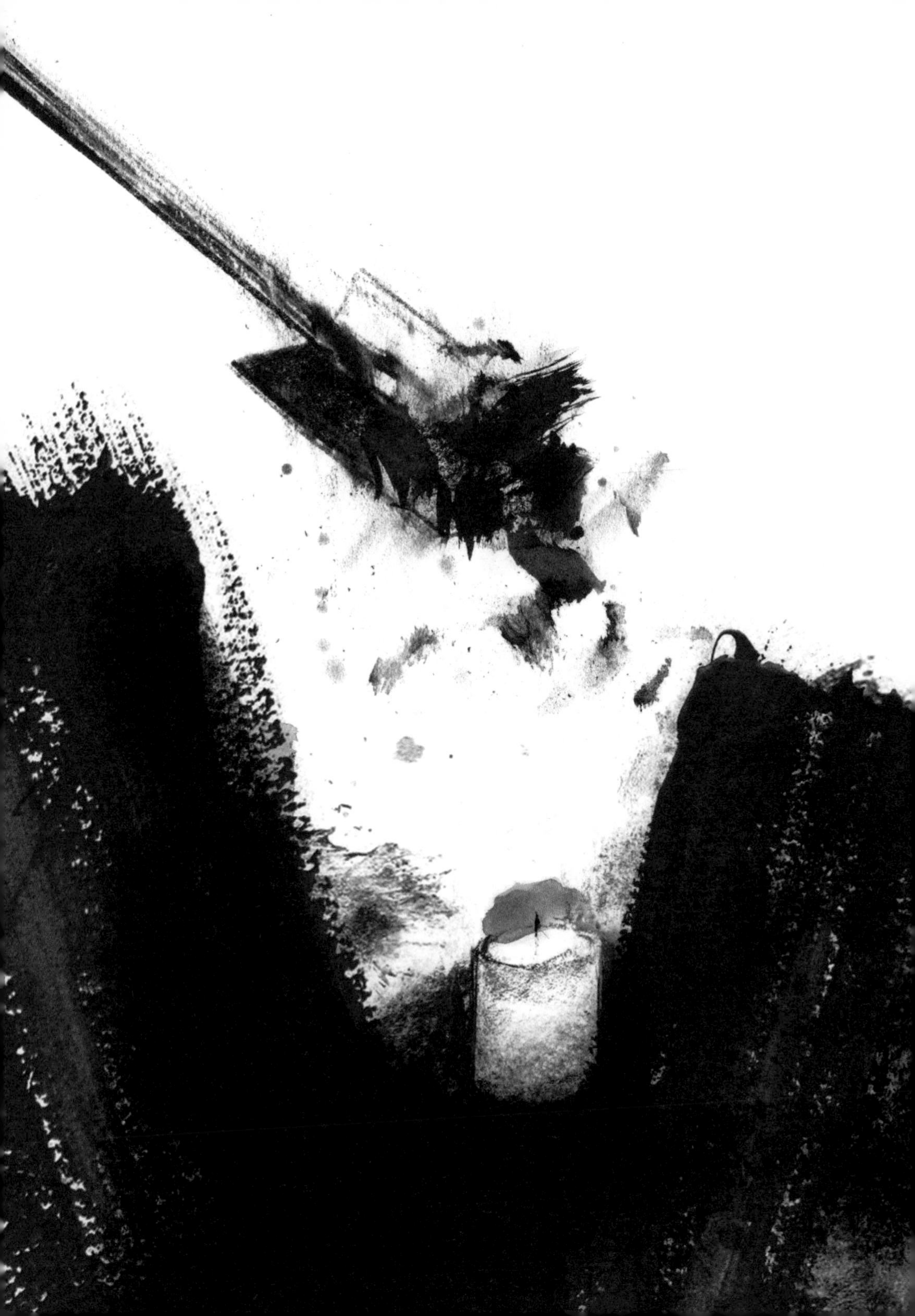

I bludgeoned the hard earth
with my spade and my sorrow.
I kissed your scruffy face.
Your paw slipped from my hand.
Curled up for the final time,
I fetched a candle and lit it
as the soil rained down.

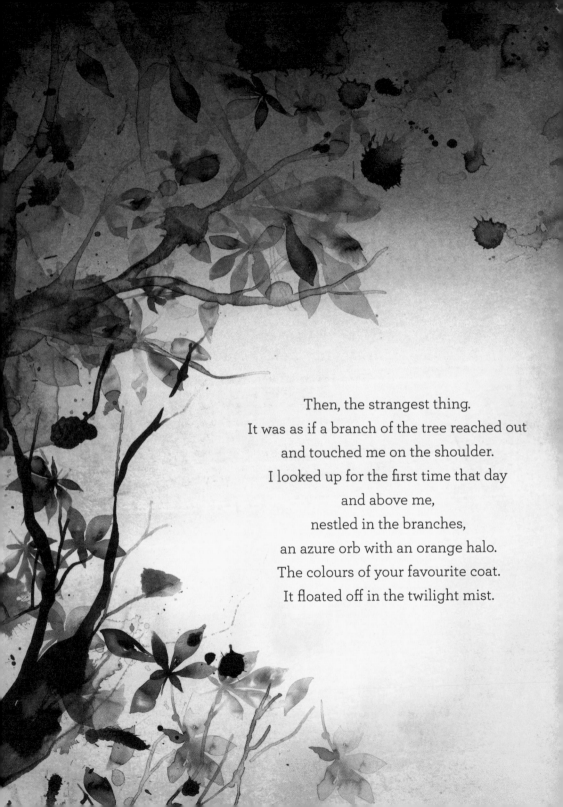

Then, the strangest thing.
It was as if a branch of the tree reached out
and touched me on the shoulder.
I looked up for the first time that day
and above me,
nestled in the branches,
an azure orb with an orange halo.
The colours of your favourite coat.
It floated off in the twilight mist.

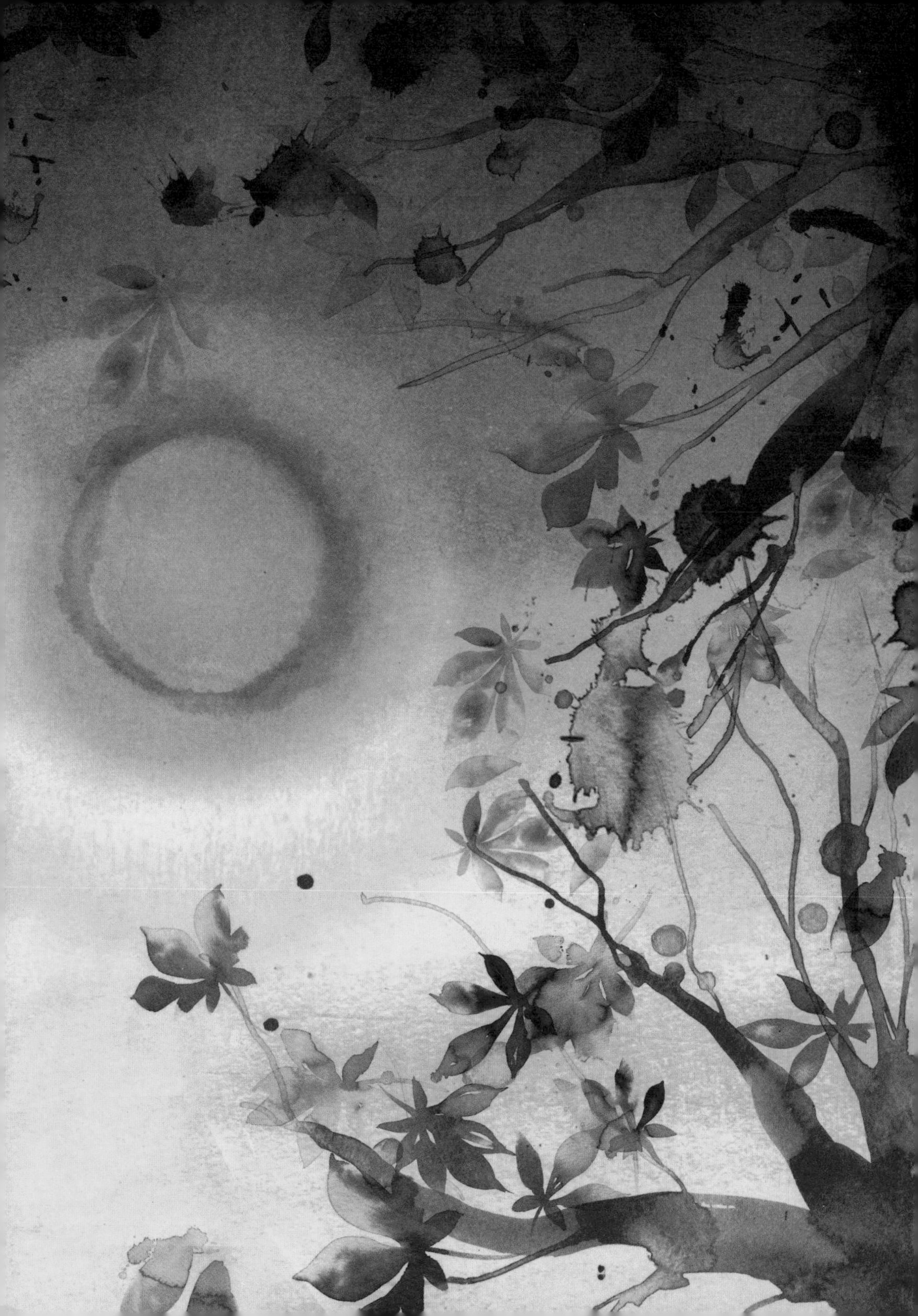

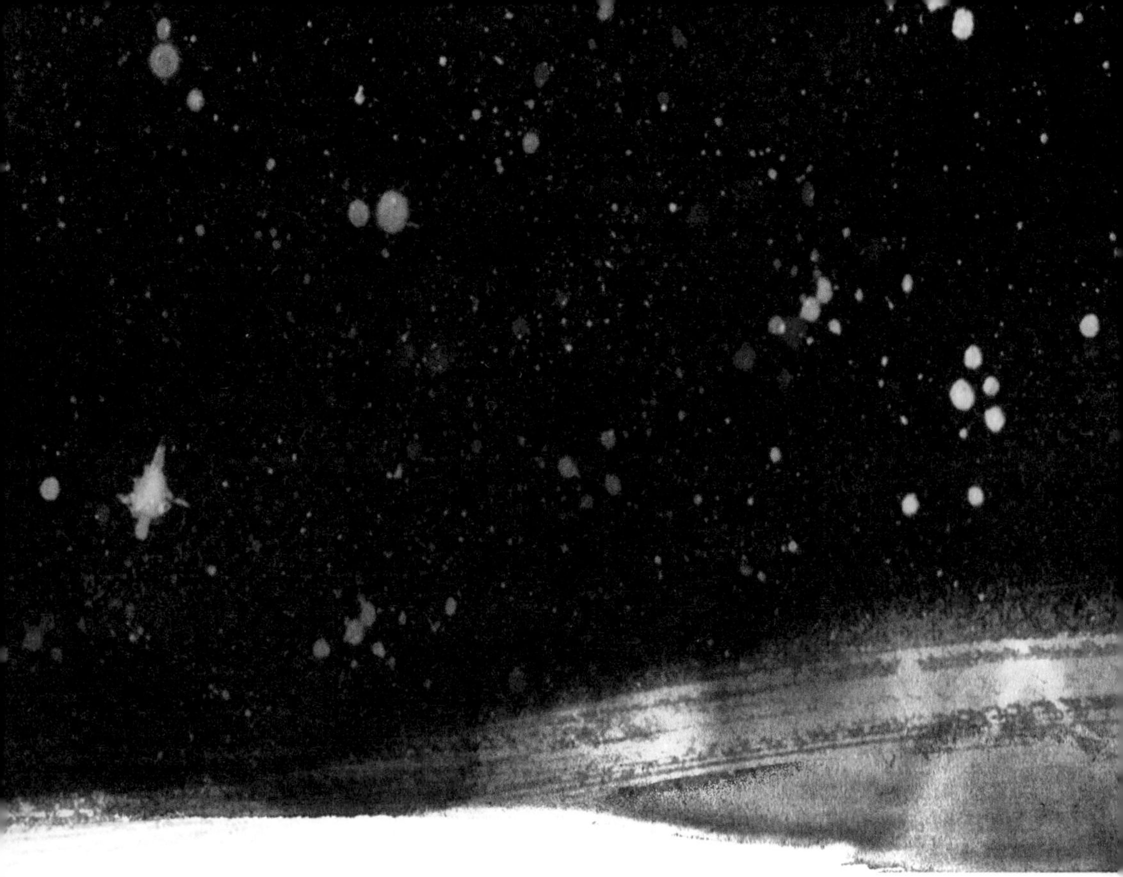

There were no clouds that night.
I found you in the sky,
running amid the stars,
like you ran through daisies,
boundless joy and infinite potential.

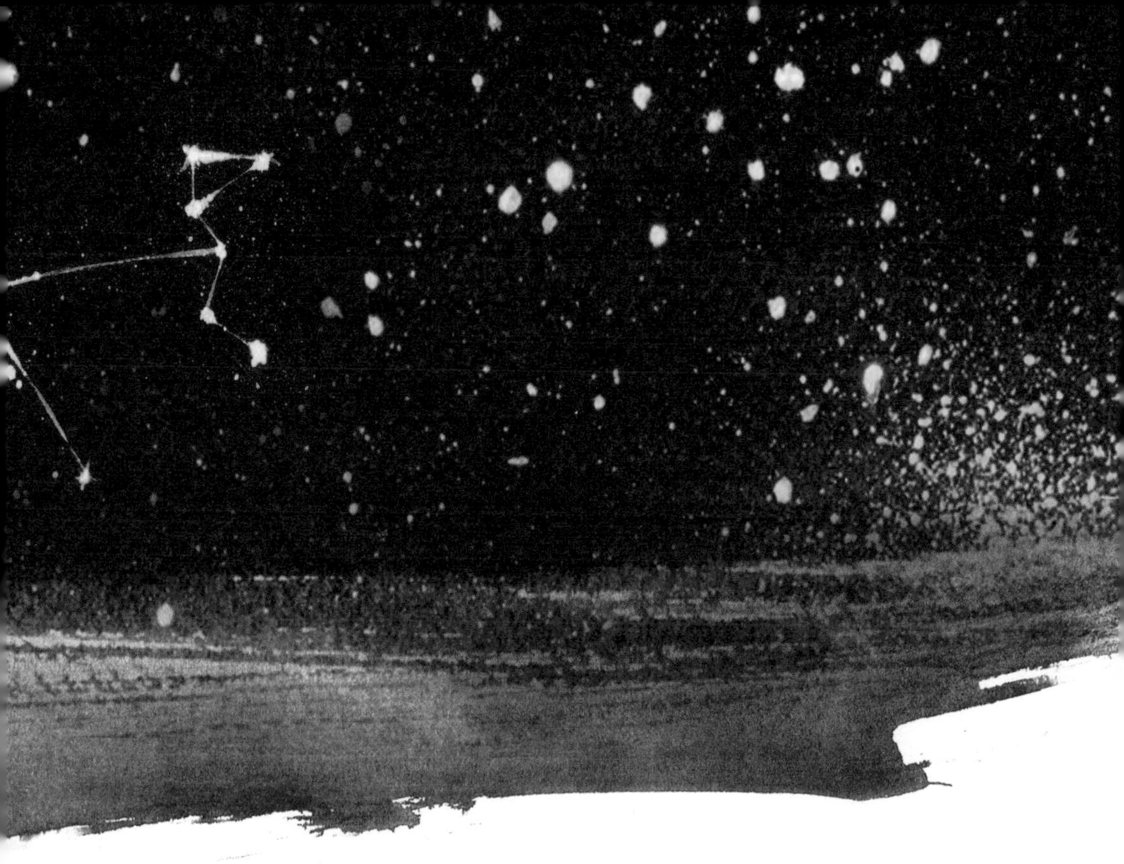

Back to the light.
Back to the brightest star in heaven.
I might wish upon a star with all my might,
but you made me realise
the light I was reaching for
was inside me all along.

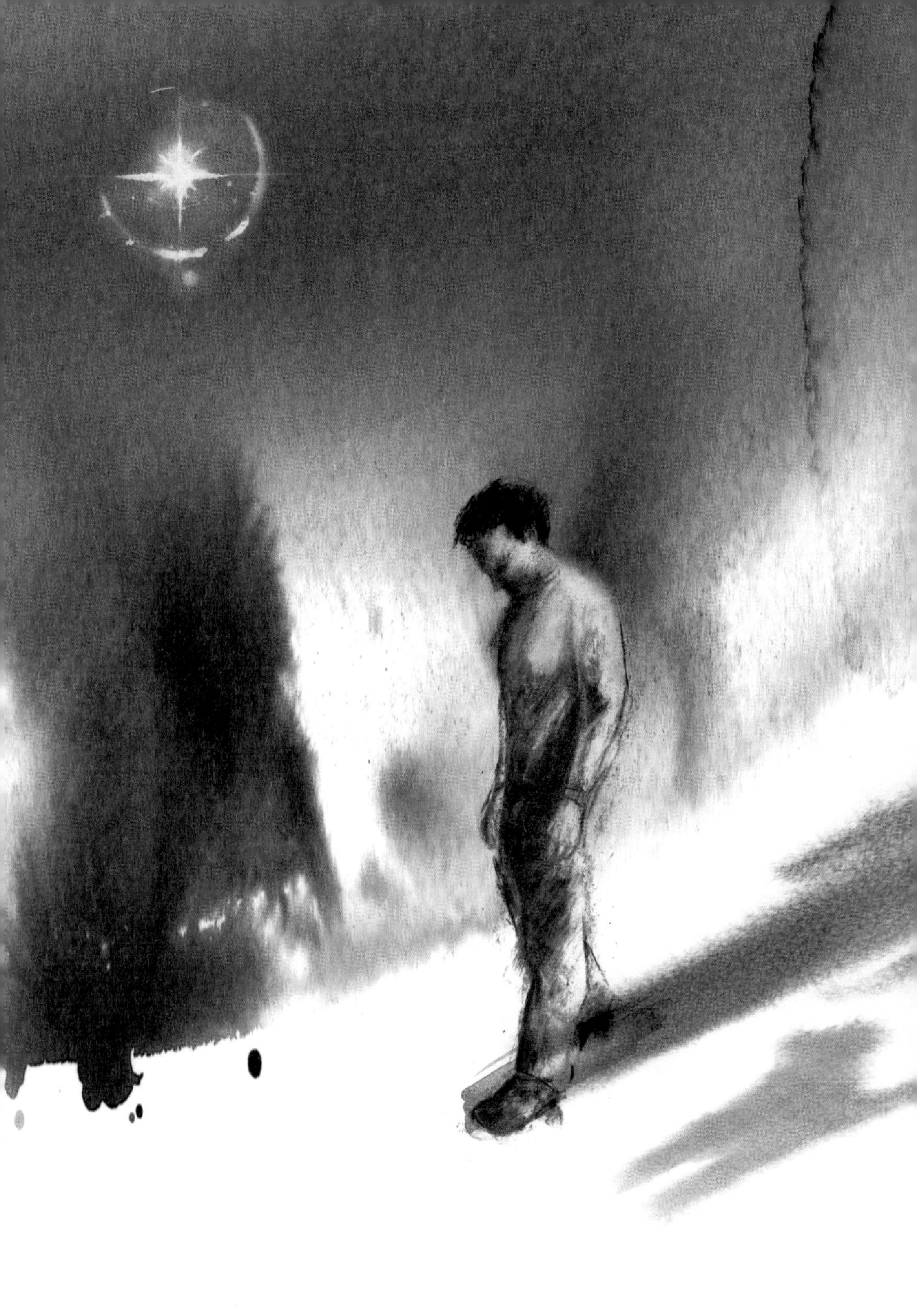

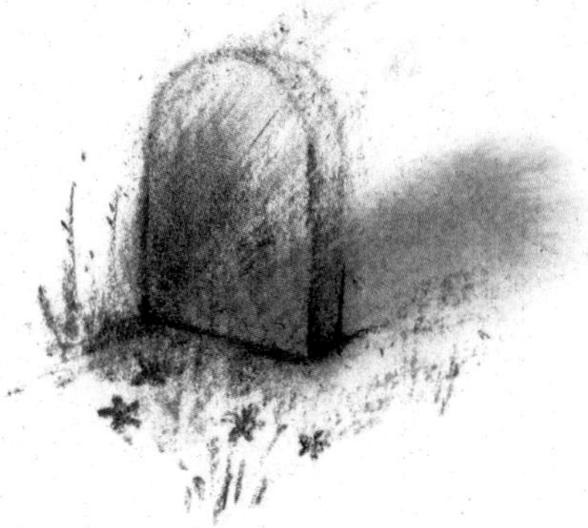

There was no darkness
you didn't brighten with your smile,
which you gifted to everyone.
But the greatest gift you gave me
was the light to find myself,
rather than accepting what people wanted me to be.

Even as I was trying to heal you,
you healed me more than medicine ever could.
I hear you whisper,
In accepting that we pass, we can be most alive.
In accepting that we lose, we can always win.

You must *forgive* the past to *give for* the present.

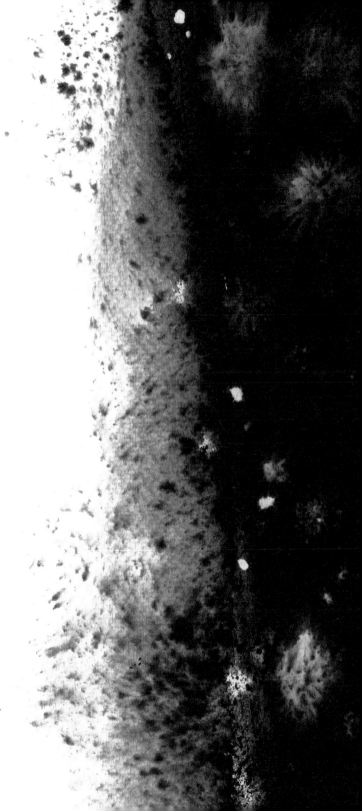

I am not scared of the
darkness any more.
The key you held opened so
much more than just a door.

A light that's always there,
even when it's not.
A love transcending
space and time.

Death is but another door
to endless possibility.
Like love without conditions,
we don't need to understand
just simply to accept.

What's meant for us
will be there for us.
For there is no fear
when all is love and love is all.
As infinite as stardust.

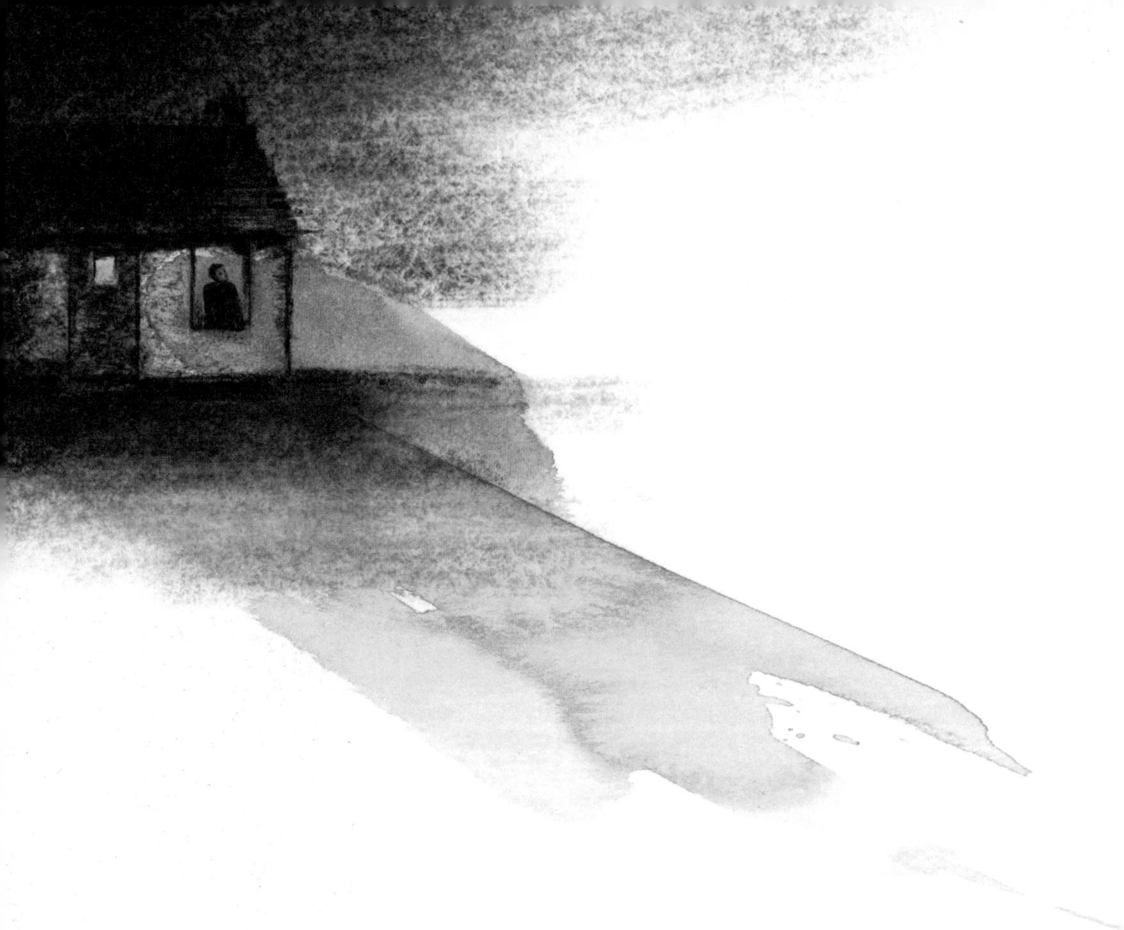

I see you daily
in the kindness of others,
in their strength and compassion,
in their giving and their forgiving.
And in their love, when it is unconditional.

For when night's pillow nuzzles back the day,
the only light you get to keep
is the love you gave away.

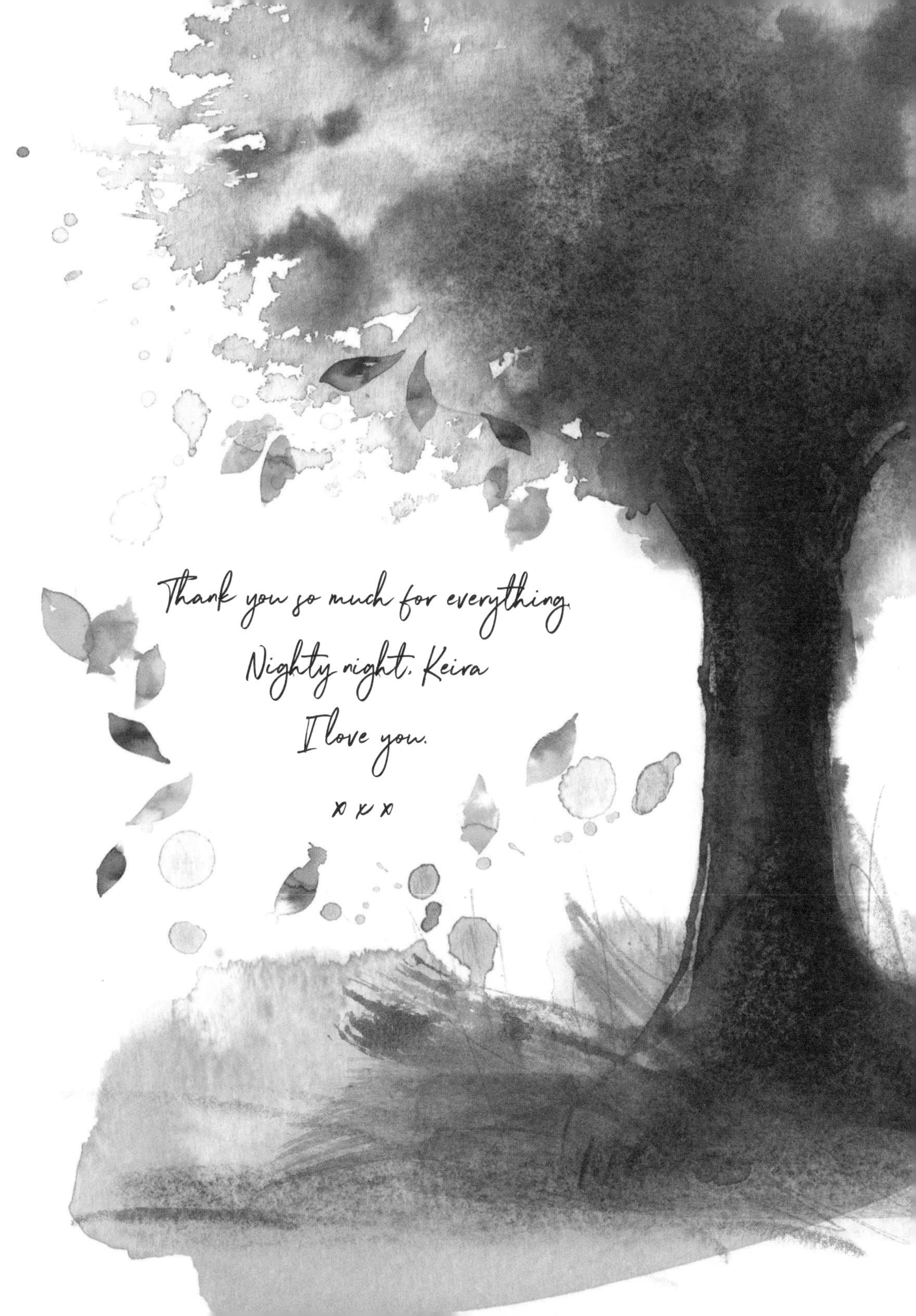

Thank you so much for everything.

Nighty night, Keira

I love you.

x x x

Afterword

Christmas morning 2022 I woke up crying. Keira was dead.
Mammy was dead. My dream to build the two greatest veterinary
hospitals in the world was dead. It seemed like everything I
cared about was drifting inexorably away and I wanted to drift
away too. I buried my head in my pillow and it was then that
Keira whispered in my ear. I kept my eyes closed and sensed her
presence as if she were actually there. I surrendered once again to
her love and let her in. That is how this book came to be.

She had been at my feet as I studied and wrote academic papers and books for what seemed like endless years. I have pushed veterinary medicine further forward in my field than anybody I can think of, and yet I am a dinosaur on the brink of extinction. Though I question everything and trust nothing, with Keira's unconditional love, I trusted everything without questioning at all. When my life unravelled, Keira's love stitched my wounds and kept me from falling apart. She will always be a unique part of me because her love is woven into me.

I see life unravel for the families of animals every day in my consulting room – when the unconditional love they share with their animal companion is perched on the precipice. To live or not to live. To try or not to try. To hold onto life or to let it slip away. This dilemma is unique to veterinary medicine, since euthanasia for humans isn't an option where I live. It's a chasm in which I have held a thousand hands and a thousand paws. People who have never cried in their lives before despite all manner of crises, cry when the animal friend they love is in pain. As one man who wept in my arms recently said to me, in reference to the dog he loved, 'She unlocked a part of my heart I didn't even know existed.'

This is unconditional love. Love without conditions. Nothing expected in return. Non-transactional. I have the honour of looking after this love every day of my life . . . a love that transcends space and time, connecting us to those we love, long after their mortal essence has passed. We don't need to understand, just surrender and accept. This is a great blessing – fear of our inevitable transience sublimated by the certain knowledge of eternal love. My love for Keira makes my heart swell right now, just as it did when her bristly smile filled my eyes with all that ever mattered.

Keira was the most patient of patients and the most patient companion I have ever had. She came through all of her medical treatments with extraordinary bravery and put up with me in all my moods and with all my faults for nearly fourteen years. Attachment to material things can sometimes smack you in the face, but only unconditional love will lick your face . . . even when you smell. I saw my own truth through her eyes. Looking in the mirror was hard . . . except when she looked in with me. Even when she was running on empty herself, she filled my emptiness willingly. She was brave enough for both of us and, though she was small, she taught me that it's the smallest things that make the biggest changes. Without each small kindness, there can be no great love.

I don't agree that 'you can't teach an old dog new tricks'. Keira and I learned from each other every day, and we changed together. We grew side-by-side. She inspired me always to keep dreaming, regardless of the setbacks. Age was never an impediment to dreaming big. Every smelly sock and soggy tennis ball was a potential dreamworld regardless of her advancing years. The night before she died, she found the candle I buried with her, when rustling for 'treasure' in a clump of packing material on the floor of my office. She chose the light to take with her, just as she has left her light behind her. Keira's heart was a battery of love, each beat radiating current into the world through every strand of her wiry fur.

It is important to acknowledge the amazing clinical teams at both my former and my current hospitals who helped save Keira's life after her horrific accident, and also the other members of her family who were present as we said goodbye to her. I co-parented Keira with my friend Amy who, at the time of Keira's arrival, worked with me as a nurse. Her son Kyle grew up with Keira and was very much her brother, and always will be. I worked so many hours in surgery, plus running the hospitals, that it worked out great for everyone. It was wonderful for Keira to have two families. With other members of the 'Fitz-family' at my practice, who loved her dearly too, we came together and laid her earthly body to rest. We held her, and each other, as Keira had held us together. Pain wracked our bodies as we prayed, thanking the universe for our little girl.

Throughout my life people have left me, as you will undoubtedly have experienced, too. Sometimes they're the people who are just passing through, and sometimes they are dear friends or colleagues, or partners who you wish you could still have in your life. For all of those I have lost, for whatever reason, I love you unconditionally anyway. I learned this from my greatest teacher – the one who never left me and never will – my little girl, Keira.

Life is a journey through darkness and light. The only way to get rid of darkness outside is to let light in. And the only way to get rid of darkness inside is to let love in. The greatest manifestation of unconditional love is when darkness isn't scary anymore. Even now, when darkness comes, she's in every speck of stardust. She was the best friend I have ever had and her spirit still guides me with the eternal light of her endless hope. From the stars and back to the stars. I can feel her light every single day . . . like the sun. The words she whispered have finally brought me peace. I hope that they bring you some comfort and light too.

With all love,
Noel x

Noel Fitzpatrick is a world-renowned neuro-orthopaedic veterinary surgeon, the founder of Fitzpatrick Referrals in Surrey, the star of the hit Channel 4 television show *The Supervet*, and no.1 bestselling author. Globally recognised for his innovative surgical solutions for animals, Noel has developed dozens of new techniques, including many world-firsts, that have provided hope where none seemed possible. Noel lives in Surrey with his Maine Coon cats Ricochet and Excalibur, and you can follow him on Facebook and Instagram @ProfessorNoelFitzpatrick, and on Twitter @ProfNoelFitz, where he would love to hear from you.

Laura McKendry is an artist-illustrator whose work combines expressive brushstrokes and decisively drawn lines to create images that convey a deep human connection with the natural world. Her original paintings are held in international private collections and she has worked with the V&A and Waterstones. Laura also teaches illustration short courses at Central Saint Martins, University of the Arts London, and creates educational content for AccessArt. She lives in London with her scruffy black dog, Bobby, and elderly tabby cat, Tilda. You can find Laura on her website www.birdandbeast.co.uk, and Instagram @birdandbeast_art.

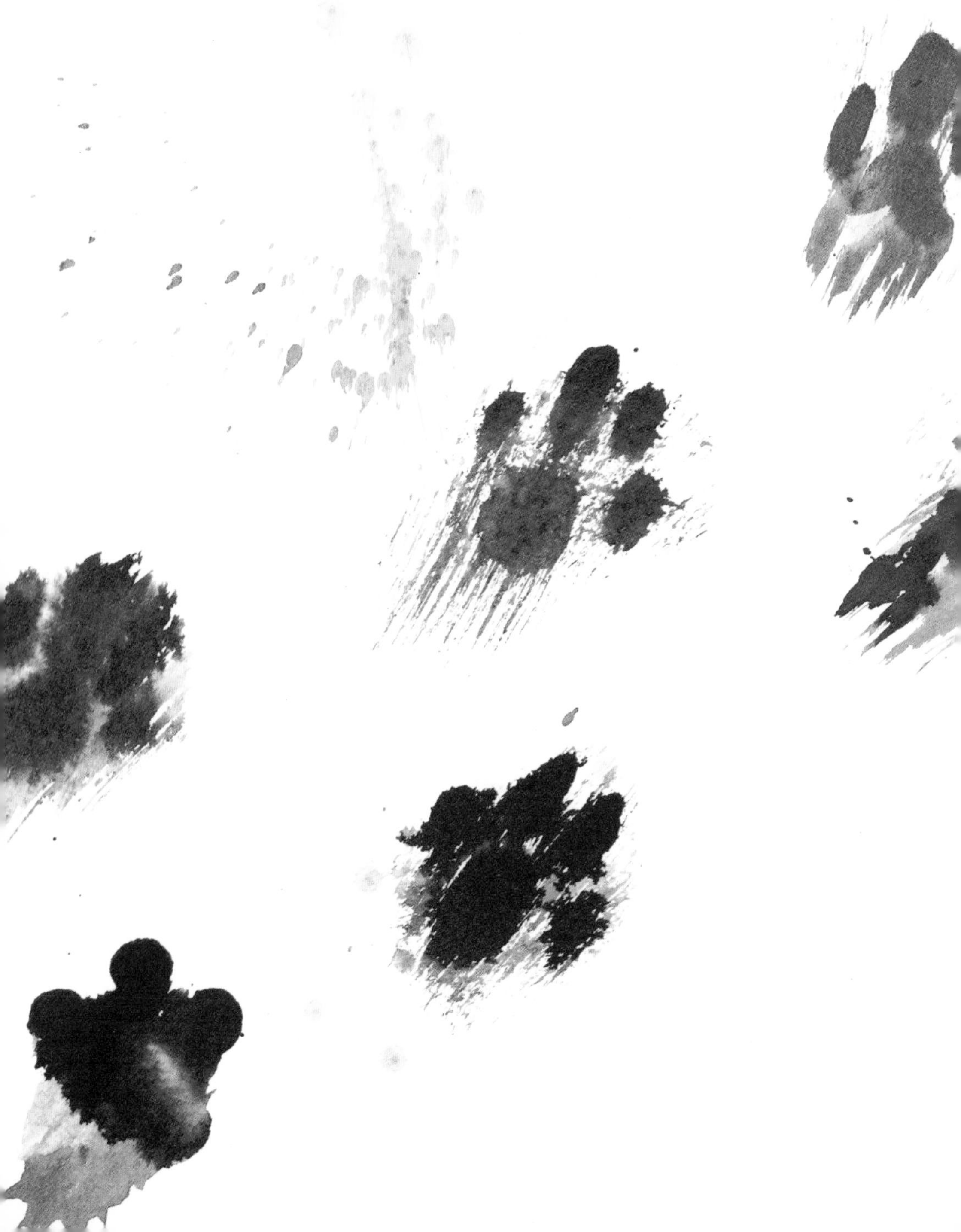